Blacksmithing
Basics for the Homestead

Drew,
Hope you enjoy
the book.
Joe De La Ronde

Blacksmithing
Basics for the Homestead

JOE DELARONDE

PHOTOGRAPHS BY JESS LEONARD

2017 Gibbs Smith Publisher

Published by
Gibbs Smith
P.O. Box 667
Layton, Utah 84041

Orders: 1.800.835.4993
www.gibbs-smith.com

Designed by Rudy J. Ramos
Printed and bound in China

Library of Congress Cataloging-in-Publication Data

DeLaRonde, Joe.
Blacksmithing basics for the homestead / Joe DeLaRonde ; illustrations by Joe DeLaRonde ; photographs by Jess Leonard. — 1st ed.
p. cm.
ISBN-13: 978-1-58685-706-6
ISBN-10: 1-58685-706-1
1. Blacksmithing. I. Title.
TT220.D43 2008
682—dc22

2008003536

WILHELM "BILLY" VOGELMAN

1903–1989

Mentor, Friend, and Old-World Craftsman

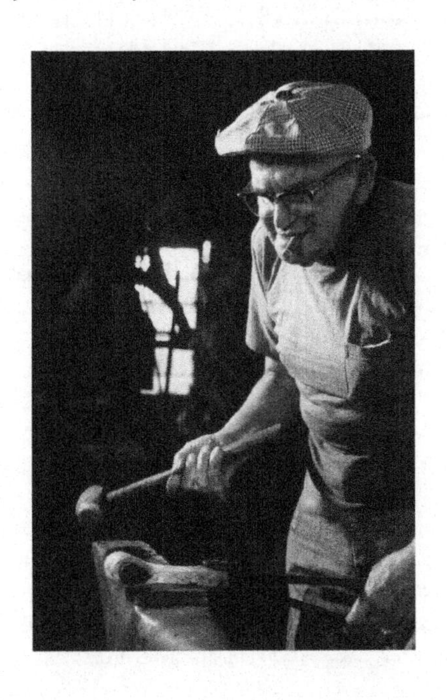

Contents

Acknowledgments

I would like to thank my wife, Marlis, for all of her support in keeping this project alive and moving forward and for all of her technical expertise with the computer and editing, without which I would have been lost.

This book could never have been written without my mentor and friend Wilhelm "Billy" Vogelman, a true old-world craftsman.

A special thank you to Jess Leonard of Jess Leonard Photography. His pictures of my work are extraordinary and I appreciate his professional skill.

Thanks also go to Mr. Bill Scurlock, of Scurlock Publishing (formerly Rebel Publishing), who was kind enough to release for use in this book excerpts from my chapter in *Book of Buckskinning IV* (1982).

Last but not least, I want to thank my parents, Gordon and Ann DeLaRonde, for encouraging me to follow my dream.

Introduction

Blacksmithing is one of the oldest and most important trades from the standpoint of its impact on civilization. It was responsible for bringing mankind out of the Stone Age. Through the years, it has changed the course of history.

Civilizations that mastered and improved on metallurgical technology in its many and varied forms tended to expand, conquer, and subjugate those that did not. The blacksmith, with his coal forge, hammer, and anvil, is the basis for modern technology.

The purpose of this book is to give you the basic blacksmithing skills to become self-sufficient on the homestead, or anywhere else for that matter. The one thing that will keep you tied to and dependent on the "settlements," so to speak, is the need for steel tools and accoutrements. Look around the average home and notice everything made of steel: cooking utensils, lighting fixtures, hinges, latches, locks, curtain rods, tables, chairs, hooks, nails, screws, pliers, screwdrivers, hammers, and the list goes on. With a few basic tools and skills, you can make the tools needed to build the homestead, from hammer, nails, chisels, hinges, and latches to amenities such as cooking utensils, lighting fixtures, wall hooks, and curtain rods.

Keep in mind that everything covered here will be done in the traditional manner—by eye. Forget pyrometers, degrees centigrade or Fahrenheit, Rockwell testing, fancy alloy steels, arc welders, and the like. You are about to enter the world of the nineteenth-century blacksmith, where skill and dedication to the craft predate mass production, assembly lines, and the computer age.

Realize that, as with any craft, it will take practice, practice, and more practice. Once you master the basics, it is just a matter of combining basic techniques in a variety of sequences to achieve the finished piece.

There are countless ways of doing any one procedure described on the following pages. Do not hesitate to experiment. You will never know it all, so be open to suggestions and be willing to learn from others. By the same token, don't hesitate to share information with others. After almost 4,000 years of ironwork, there are few "secrets."

1 History of Blacksmithing

FACTS, FANTASY, AND OLD WIVES' TALES

Next to glassblowing, woodworking, and pottery, blacksmithing is one of the oldest trades in the world. It is also one of the most important and demanding. The blacksmith was called the craftsman's craftsman since he made the tools for the other trades.

Blacksmithing, as near as can be determined, originated in the Caucasus Mountains about 4,000 years ago. As it expanded and became a primary trade, guilds were formed to maintain quality and guard trade secrets. Prior to the 1800s, virtually all trades, including blacksmiths, passed on their training and secrets through the apprenticeship method; there were no textbooks. Very little, if any, written information was available, except perhaps notes and techniques written down by individual smiths. As a result, smiths tended to shroud their trade and individual secrets in mystery, thereby keeping the uninitiated and curiosity seekers from learning some of their more highly prized "secrets." It was just good business. If people thought you had a special way of tempering clock springs, you would garner the majority of that type of work in your area. Thus, with no written record, many trade secrets have been lost forever. These secrets were not passed on, for whatever reason, and are now lost.

I worked under William "Billy" Vogelman, a tenth-generation blacksmith from Emmetsweiler, Germany. He mentioned that his grandfather knew a way to work bronze so it would hold an edge the same way as steel. That "trade secret" has been lost as well.

Here's an example of how a modern-day "secret" started, regarding Billy's technique for sharpening plow lays. Years ago in the early and mid-1930s, the dust bowl hit the central plains states, including Iroquois, South Dakota, where Billy had his shop. During this time, virtually all the topsoil disappeared, leaving in its wake clay, gravel, or sandy subsoil. On weekends, Billy would drive around the countryside and note where sandy spots or heavy or light clay spots were and which of his customers farmed the various types of soil. By knowing the type of soil, Billy would then know how much "suck" to put on the plow lay point to keep it in the ground. The other thing that Billy did was to polish the lay in the same direction that the dirt would roll off; thus the lay always scoured clean as soon as it was in the ground. These two factors set the stage for one of Billy's "secrets." Since plow lay work was one of the mainstays of the rural blacksmith in those days, and since Billy was only a one-man shop, it was inevitable that he would work late into the night, especially during the spring and fall plowing. It wasn't long before Billy started having customers show up from areas farther and farther away because they had heard about his "secret" method of sharpening plow lays so they would stay in the ground, cut, and scour clean. The combination of knowing the ground, polishing in the right direction, and, of course, working late at night (to hide his secret, or so they thought), created the "secret methods" and the aura of mystery surrounding his technique. Billy never said he had a secret method for sharpening plow lays, but then he didn't dispel the notion either. After all, he had the lion's share of plow lay work in his area. To this day some of the old-timers still talk of Billy Vogelman's "secret" for sharpening plow lays.

When blacksmithing came to the New World, distances between settlements made the guild system impractical. The blacksmith in the New World could not afford to specialize, so he became cutler, farrier, and toolmaker, creating anything made of iron.

Although the blacksmith may have specialized in a certain area based on the old-world guild system, he soon adapted to his new environment and became a master tradesman in the truest sense of the word.

Following World War II, the blacksmithing trade was almost lost, being replaced by the welding and machine shop that we know today.

During the late '60s and early '70s, there was a renewed interest in many of the old trades—among them, blacksmithing. As a result, the trade did not die but flourished and is still alive and well in the twenty-first century.

The aura of mystery surrounding blacksmithing is inherent because it is a demanding, complicated craft that even today the layman knows very little about. The only reason I would dispel some of the myths surrounding blacksmithing is to save you wasted time and energy. For example: "Coke is the only fuel you can use in the making of Damascus steel." Bunk! Damascus was made using charcoal long before coal became the standard fuel for forges. True Damascus is created during the smelting process. Modern-day Damascus is actually "pattern-welded steel" where several types of steel, i.e., mild steel, high-carbon steel, nickel, and others are forge-welded together and then twisted and otherwise manipulated to create various patterns. I have seen some smiths run two forges in their shops, one that they are working at, the other going just to make coke. Since the forge you are working is coking coal as it burns, why run another one? Once again, if it sounds complicated and time consuming, not everyone will want to do it, which will thus cut down on the competition.

Another myth is, "Knives must always be quenched with the blade running north and south." Pig pucky! I have forged and quenched hundreds of knives, north, south, east, and west and have found no difference! It's like some of these "secret" tempering solutions for knives that you hear about: wing of bat, eye of newt, left wing primary feather from a raven that has flown north to Stonehenge on the third hour of the summer solstice! The only secret is how people can be made to believe all that. Blacksmithing is based on common sense and dealing with basic physical properties of the materials used: steel, heat, and water. Mastering these takes enough skill and practice without cluttering it up with nonsensical gibberish. These are the kind of things that prompted me to write this book. I was fortunate enough to apprentice under an old-world master. That is where this information comes from. I am just passing it on to the next generation.

2 Setting Up Your Shop

SHOP SETUP

Shop setup should be tailored to your needs and what you feel comfortable with. However, there are several basic things common to most shops that should be kept in mind. The forge is the heart of any shop. It is normally set up against a wall; however, make sure you keep the other three sides open with plenty of room to maneuver larger pieces you are working with. Some work may require a person on each end to turn the piece in the fire and maneuver it onto the anvil.

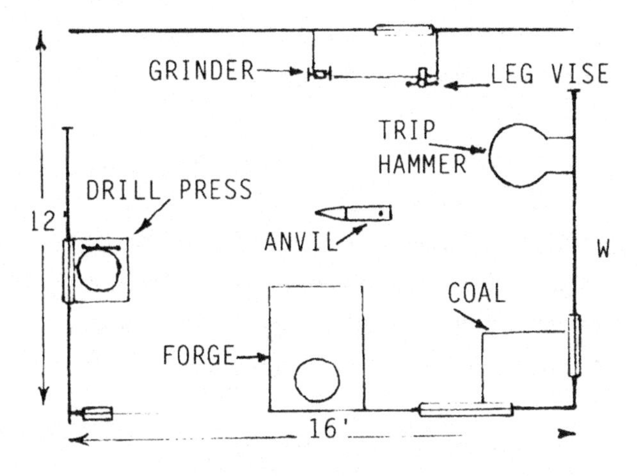

Figure 2.1 (facing). The author's shop.
Figure 2.2 (above). Layout of the author's shop.

As with any craft, a dedication to practice, practice, and more practice is not only desirable but essential to becoming proficient. Reading books and collecting tools won't do you much good until you start lighting the forge and practicing. We all have to start somewhere. If possible, find a good smith to watch or visit. I stress "good" because there are a lot of people who have no business having a hammer in their hand much less being called a blacksmith. A bad smith will do you more harm than good. However, a day spent watching a competent blacksmith at work is worth a month of reading!

Remember this: the frustration factor will increase in proportion to the number of burned fingers and misplaced hammer blows. This becomes a geometric progression when you begin forge welding. But then, to quote an old sage, "The man what ain't never failed, never did nothin'." Heating, shaping, and tempering are dependent upon visual recognition coupled with timed physical activity to achieve the end result. This will only come about with practice and familiarity with all aspects of the forging process. It will not come to you overnight. You must be willing to practice and learn from your mistakes. In time, the feel will come naturally.

As with any trade or craft, you must obtain certain necessary basic tools and equipment. You can cut corners and try to save money with half measures, but there are two items that require a substantial cash outlay and in the long run will be worth every penny: the anvil and the forge. Do not scrimp. You get what you pay for. A list of suppliers can be found in the resource section.

There are three tools basic to all forge work: the forge, the hammer, and the anvil. With these three pieces of equipment, you can make anything else needed. Remember, one of the interesting aspects of blacksmithing is the ability to make all of your own tools. Since many of the tools are readily available through various suppliers, it might be more practical to obtain them by purchasing them rather than making them all from scratch. Another method of obtaining the tools is through farm and ranch auctions, flea markets, online auctions, and scrap yards.

THE FORGE

A good new modern-day forge will run anywhere from several hundred dollars to a thousand or more. However, the old forges can still be found at farm and ranch auctions, particularly in the upper Midwest and Great Plains states. You may also choose to make the old-style double-action bellows (Figure 2.3) and forge. In the early days, a large double-action bellows was used. Double-action means that the bellows blows a continuous air blast even when filling with air. There is an excellent little book by Robert M. Heath listed in the resource section that will give you step-by-step instructions. Keep in mind that these are the coal-fired forges. We won't be discussing the propane or natural gas forges since we are talking about homesteading and being self-reliant.

The forge is merely a device for holding the fire and blowing air up through it to maximize the heating ability. It is composed of three basic parts: the table; the tuyere or fire pot; and a means of blowing air, usually by a fan or bellows. A small forge will work fine, but if you have room for a large setup—3 x 4 feet or larger—it will work much better.

Forges come in varying sizes, from large permanent forges made of brick found in blacksmith shops to the small portable riveter's and farrier's forges found on farms and ranches.

You can jury-rig any number of different types of forges. However, you will be money ahead if you

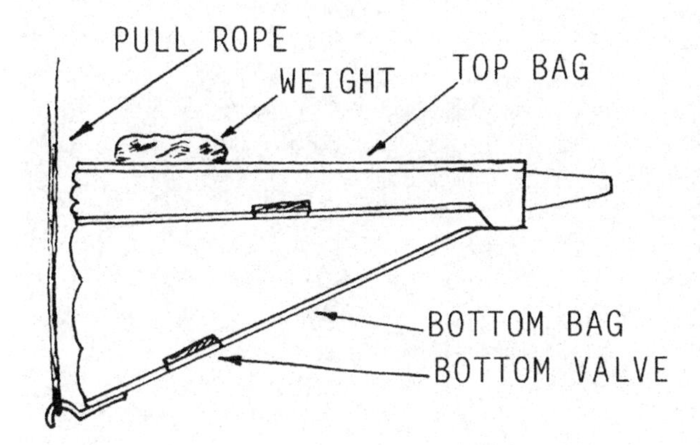

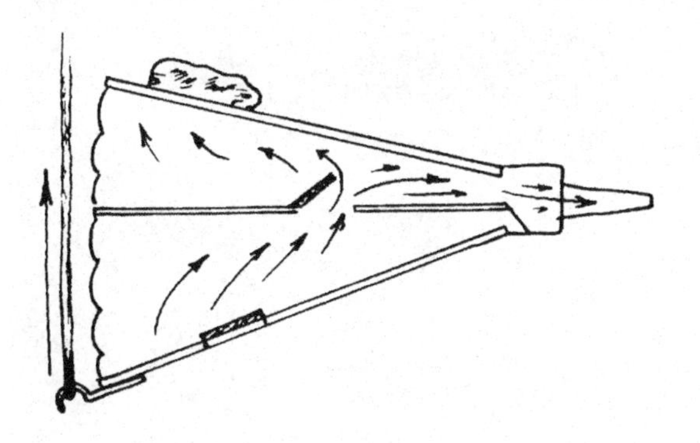

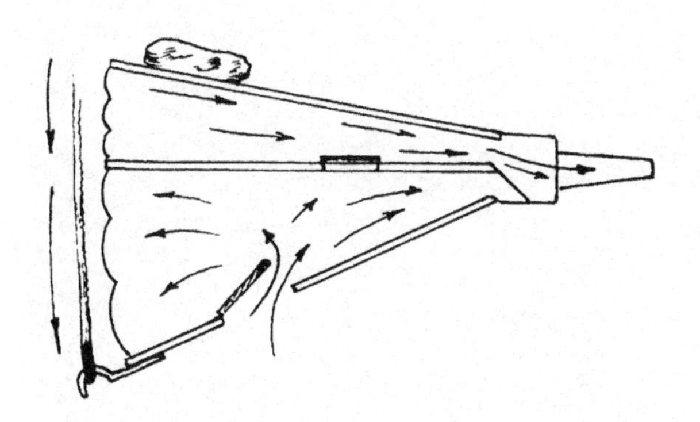

Figure 2.3. Double-action bellows.

invest in a good riveter's or farrier's forge or, better yet, build a good permanent forge. Many books listed in the resource section give detailed plans for the construction of a forge.

THE ANVIL

If the forge is the heart of any shop, the anvil and hammers are the brain, because it is with these that the actual forming and shaping is done.

The anvil will be an expense, but as with the forge, don't scrimp. Expect to pay $2 to $3 a pound or more for a good new anvil. A good old anvil, which again can be found at farm and ranch auctions, will usually run $1 to $2 a pound. These are getting harder and harder to find, though. There are two types of construction for anvils, depending on how they are made. The old style is of a two-piece construction: a cast-steel body with a high-carbon-steel face forge welded to it. The modern anvils are cast of high-carbon steel all in one piece, with only the face being tempered.

A good test for an anvil is to tap it with a hammer. If it has a good ring to it and the hammer bounces, it is a pretty good bet you have got a good anvil. A "dead" anvil, which will not have a ring or bounce, should be avoided—you'll be working against yourself.

There are two basic styles of anvil: the "English" style, which is the most common in the United States, and the "shoeing" anvil, which is used by farriers (horseshoers) and is basically a modified version of the English.

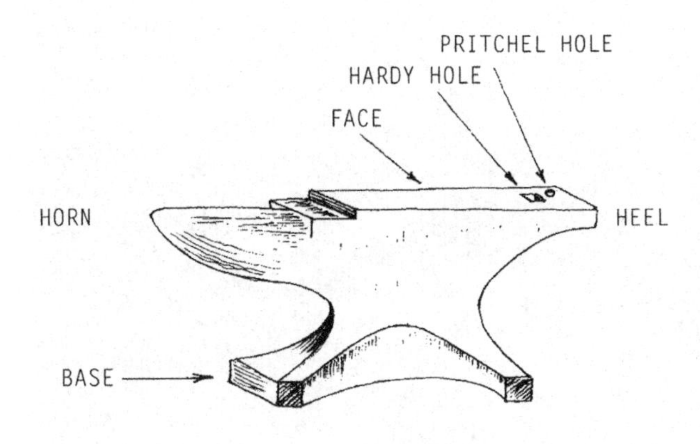

Figure 2.4 (facing). The author's anvil, with various hardy tools arranged on the anvil block.

Figure 2.5 (above). Parts of the English-style anvil.

A good size for a shop anvil is 100 to 150 pounds. (This will work for anything you will need to do around the homestead.) I have a 170-pound shoer's anvil that has served me well for more than thirty years, and a smaller 80-pound Trenton anvil I use when doing demonstrations and workshops on the road. To determine the proper height for the anvil, stand straight with your arms at your sides and make a fist. The top of the anvil should just touch your knuckles.

The hardy hole, which is the square hole near the heel of the anvil (see Figure 2.5), is for the use of not only the hardy—a vertical chisel used for cutting stock—but also myriad other specialized tools such as fullers and scroll forks, and for bending and forming tools of all types. The pritchel hole is a small round hole near the heel of the anvil (see Figure 2.5) and is used for punching, bending, and heading.

An interesting side note is that many ships coming from Europe to the New World used anvils as ballast in the bottom of the ships. Upon arrival, they were off-loaded to be replaced with quarried granite and marble block to take back to Europe.

THE HAMMER

For blacksmithing, there are basically three types or styles of hammers used. The most common is the cross-peen, followed by the straight-peen, and to a lesser degree, the ball-peen, often referred to as a "machinist" hammer. The straight-peen hammer is used more by farriers, although I have used it on occasion for my work in general blacksmithing.

I use three sizes of cross-peen hammers: an 8 pound, a 3 pound, and a small 3/4 pound. (See Figure 2.6 on page 22.) A word of caution here: I do not advise starting out with the 8-pound hammer. Work up to it gradually. That's assuming you want to use one that heavy. A good 3-pound cross- peen hammer will work for most of the projects that you will be doing. It is easily controlled and won't tire you. Remember that hammer control and accuracy are more important than power or brute strength. Find a weight you are comfortable with and that is easily controlled when starting out. Blacksmithing is a demanding trade, not only mentally but physically. It will take time and practice to work up to the heavier hammers and maintain control.

Figure 2.6. From left, 3/4-pound, 3-pound, and 8-pound cross-peen hammers.

FORGE AND ANVIL TOOLS

To work the fire properly, three basic tools are required: a paddle shovel, a hooked poker, and a straight poker. (Figure 2.7.) These should be the first three items made because you won't find them ready-made from any supplier that I know. You should also make each with a different style handle. The reason for this is that you feel rather than see the tool, which is important as the projects become more involved. It is better to know which tool you have by feel than to have to take your eyes off a small piece of iron in the fire and lose it because you had to visually check to see which tool you were picking up.

One tool I keep handy is a water dipper, which I use to keep my coal wetted down. It is really nothing more than a container the size of a 10-ounce soup can with a long handle attached.

The hardy is a cutting tool used in conjunction with the anvil and many blacksmithing tools. It comes in various shapes and sizes as shown in Figure 2.8.

A variation of the hardy is the fuller, which is a blunted hardy used for drawing out, rough-forging a tenon shoulder, and other jobs. (See Figure 2.9 on page 24.)

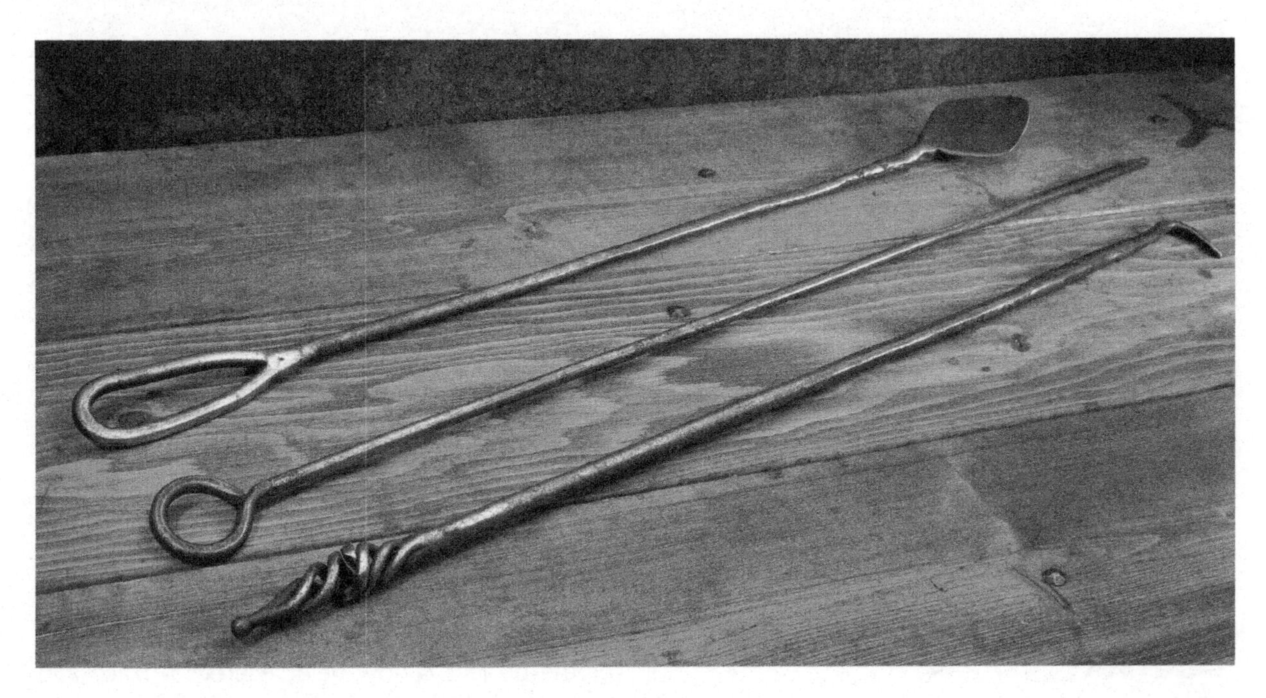

Figure 2.7. Paddle shovel (top), straight poker, and hooked poker.

The holdfast does just what its name implies: it holds fast. This same tool is used by woodworkers on their workbenches for pieces that need to be planed or otherwise worked. For the blacksmith working by himself, it is one of the handiest tools available because it becomes that third hand required for so many projects. It is particularly handy when splitting or punching. Some of the holdfasts I've seen are used in the pritchel hole and are very lightweight: $^1/_2$- to $^3/_8$-inch round stock. The one I use, and that has worked very well for me, fits the hardy hole—which is $1\,^1/_8$-inch square—and is made of $^3/_4 \times 1$-inch rectangular stock, 20 inches long. Make sure your holdfast stock is smaller than the size of the hardy hole. (Figure 2.11.)

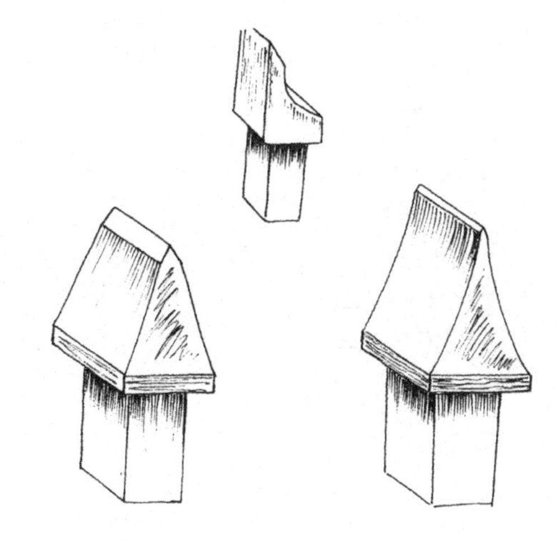

Figure 2.8. Three types of hardies.

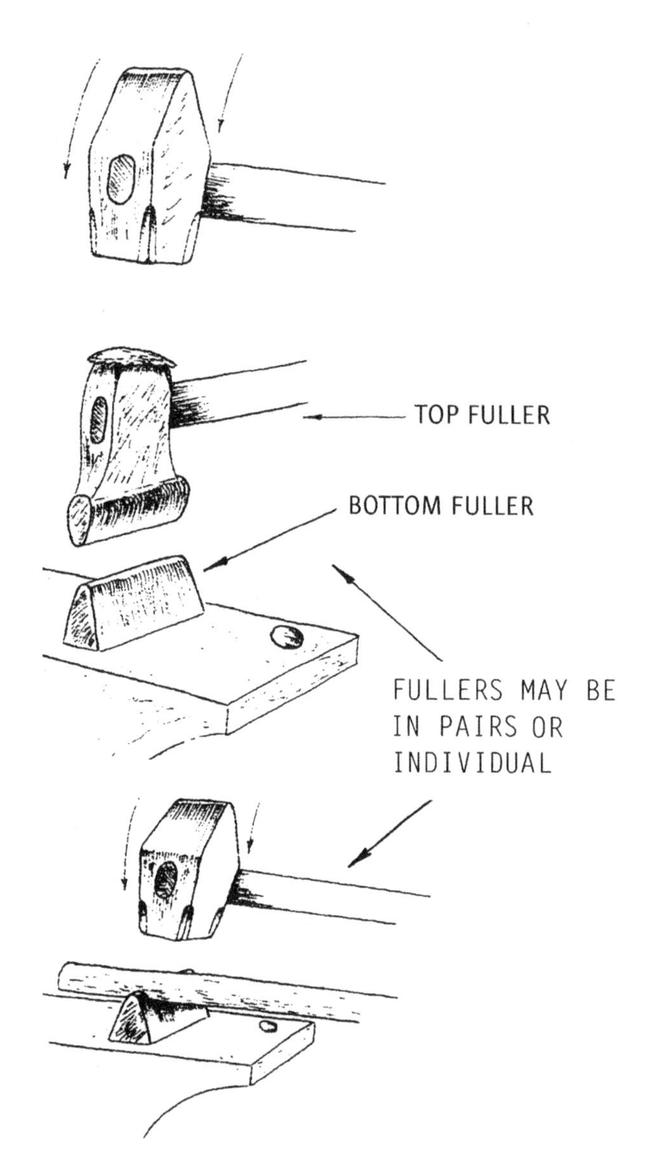

Figure 2.9. Fullers.

The soft block is usually used in conjunction with the holdfast. It is a piece of mild (soft) steel about the same width as the anvil and about 6 to 8 inches long. When something is being cut off or split, it is done on the soft block. It keeps the chisel from cutting through to the hard surface of the anvil, thus dulling its cutting edge. (See Figure 2.11.)

To use the holdfast and soft block together, lay the soft block on the anvil as shown in Figure 2.11 and place the holdfast loosely in the hardy hole. Now, take the red-hot iron to be split (or cut off), lift the holdfast, place the iron on the soft block, put the holdfast down on top of it and tap the holdfast with your hammer to set it. Your stock for the project is now held securely in place and can be cut or split. The above approach must be done quickly and smoothly so your splitting or cutting may be done before the piece cools. On large pieces it may take several heats. You will begin to appreciate the value of a good apprentice after fumbling around with all these tools the first few times. Of course, if you had an apprentice, you wouldn't need the holdfast.

The swage block is made of cast iron and comes in many different shapes and sizes. (Figure 2.10.) Some specifically used for the forging of rifle barrels are relatively small and weigh only 50 to 80 pounds. The large general-purpose blocks weigh up to 200 pounds or more.

The primary purpose of a swage block is that of a forming die. While the metal is hot, it may be

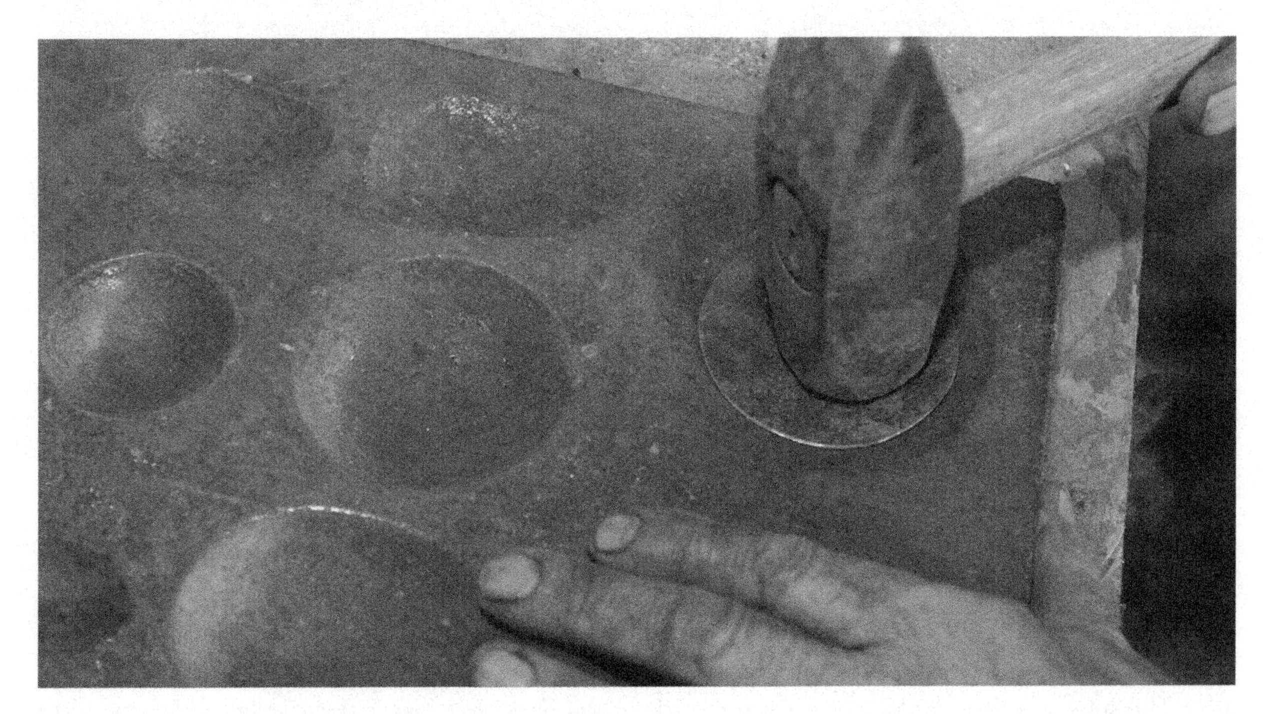

Figure 2.10. A candleholder drip pan is formed using the swage block.

pounded into any of the numerous indentations in the block and "formed." They are really helpful when making the rolled-handle "sockets" like those used on chisels, lance points, shovels, and spades. The block pictured here is very old and has seen considerable use.

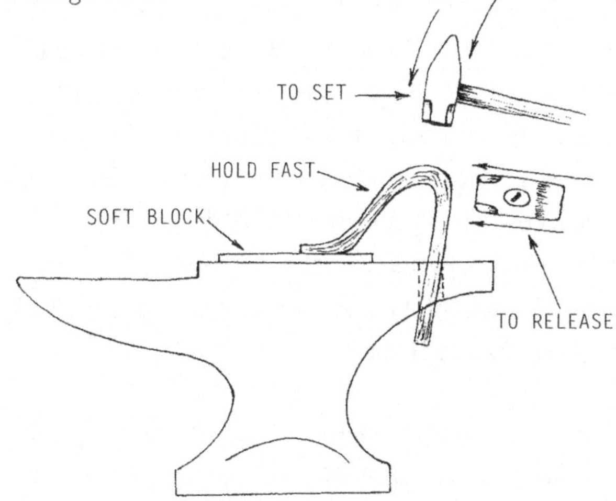

Figure 2.11. How to use the holdfast.

The bick is a small, tapered, cone-shaped tool normally used in the hardy hole. It may be vertical or horizontal. (See Figure 2.12 on page 26.)

Coal can be a problem, depending on where you live. In years past, when there was a blacksmith in every town, blacksmithing or coking coal was available from many sources. With the disappearance of

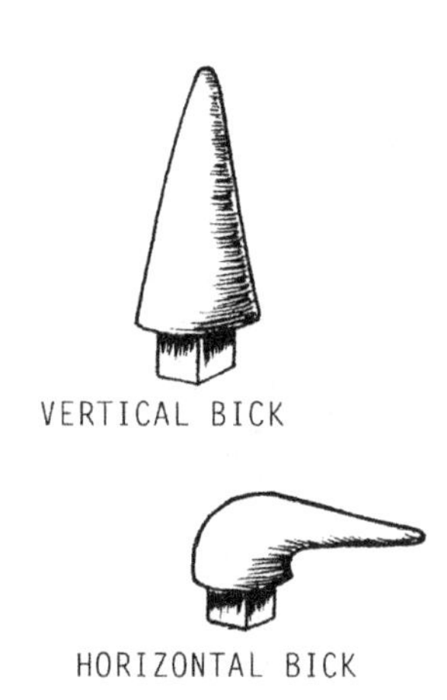

Figure 2.12. Two types of bicks.

the blacksmith, that type of coal became a specialty item. The substance you are looking for is referred to as blacksmithing coal, coking coal, metallurgical grade coal or met coal, Pocahantas or Pocahontas pea coal. It is soft, low sulphur, and clean burning. The best coal is found in the East. However, deposits of good coking coal are also found in the West, in southeastern Oklahoma, northern Arkansas, Colorado, northern New Mexico, and around central Utah. See the list of coal suppliers in the resource section.

Charcoal will work well, but it is a slower heat and is expensive. Because of its slower heat, it works well when forging fine, high-carbon cutting tools such as scalpels and small knife blades.

Flux is one item you will need for forge-welding. It may be ordered through supply houses dealing in blacksmithing, farrier tools, and equipment. The flux I have used for years with excellent results is anhydrous borax. It is cheap and readily available at chemical supply stores. See the resource section for suppliers. Billy once mentioned that, in years past, his father used good clean river sand. I've found that silica sand, which is used in glassmaking and sandblasting, will work as a flux. When the metal is at a red heat, sprinkle the flux at the point of the weld. You will notice that it melts and covers the surface of the two pieces that are to be welded. The flux acts as an oxidation retardant, i.e. it keeps oxygen out of the weld which inhibits oxidation of the steel in turn allowing a solid weld.

Steel is easily found by contacting suppliers of steel and scrap yards. Hot rolled, mild steel will work for everything but springs and cutting tools. High-carbon steels may be obtained from scrap yards, farm junk piles, or specialty houses.

Don't even mess with the modern alloys. They will bring you nothing but frustration. Good sources of high-carbon steel are coil springs, leaf springs, horseshoe rasps, and worn-out files. The old black iron or wrought iron is still around in limited quantities but is hard to find. It has a fibrous grain similar to wood. It resists corrosion very well, although it must be worked at a very high heat. Steels can be confusing to the uninitiated, which may explain why some smiths, especially those who make knives, like to throw around numbers like 1095 or 1131. I guess it is supposed to impress the neophyte. I once had one of these people ask me what I made my Spanish Belduque out of. I replied F150CS, and he gave a knowing smile and walked off mumbling something. A friend of mine came over and said he'd overhead the conversation and wondered what F150CS was. I told him it was a Ford F150 coil spring, which is what I use for that particular style of knife. Exotic alloys are fine if you have the equipment and expertise to heat treat them. For what we will be covering here, stay with the water-quench and oil-quench high-carbon steels and you should have no problems. This will all be covered in more detail in the chapter on tempering.

ACCESSORY TOOLS

A good leg vise (Figure 2.13) should also be a part of your shop equipment. New ones are not cheap, but good used ones can still be found at farm and ranch auctions, junk shops, and antique shops at fairly reasonable prices. A mechanic's vise will work in a pinch, but it will need to be the heavy-duty variety.

Figure 2.13. Leg vise.

A good drill press should also be a part of your shop equipment.

If you have the chance to get a 25- or 50-pound trip-hammer, by all means grab it! They are worth their weight in gold.

There are several types of trip-hammers: mechanical, compressed air, and hydraulic. The most common and easiest to maintain is the mechanical type. Because it can be run by electricity, water-power, or gas engine, it is the only one we will deal

Figure 2.14. A 25-pound Little Giant mechanical trip-hammer.

Figure 2.15. Hot cutter.

with here since it lends itself to the homestead environment. The most common of the mechanical types is the Little Giant. (See Figure 2.14.) They can still be found once in a while at farm and ranch auctions, junk shops, or junkyards. However, since they have gained popularity they have become increasingly more expensive and harder to find. Expect to pay $1,000 or more for a good reconditioned hammer.

The 25- or 50-pound hammers are best for most general work. Anything larger than that is overkill. The large 100 pound, 500 pound, and up are reserved for rail yards, shipyards, and other heavy industry.

One of the handiest pieces of equipment I have is the hot cutter (Figure 2.15), which is homemade. I do not know of any commercially made hot cutters available. As the name implies, it is used for cutting hot iron. I find it especially useful for cutting the edge shape on my axes and tomahawks. It also works well for cold cutting 16- to 12-gauge lightweight sheet metal.

3 Shop and Safety Tips

As with any pursuit, there are some basic safety procedures necessary to follow, and tips that, based on my experience, may make things easier.

- Wear eye protection at all times. This cannot be stressed enough. It is a cliché, but you only have one pair of eyes. Protect them!

- Consider everything around the anvil as *hot.*

- Do *not* use gloves. You will get used to picking up hot items with gloves and someday you'll do the same without them. You get the picture. You also lose tactile sensation by wearing gloves, and you need that sense to know *how* you are striking the iron and anvil. Gloves *may* be used on occasion to protect your hands while splitting or punching larger pieces. Heavy cotton gloves, sometimes referred to as "chore" gloves since they are used on farms and ranches, work best. Stay away from leather.

- When checking to see if a piece of metal is hot before you pick it up, use the back of your hand. If it is hot and you do burn yourself, you can still hold onto things and continue working.

- When you want to measure the circumference of something that needs to be collared or sleeved, use a piece of long grass. It's pliable, it won't kink or stretch, and it's plentiful.

- Remember that, in addition to being used in the shaping of steel, your anvil can also be used as a measuring tool. (See Figure 3.2.) Memorize the respective distances on your anvil—horn length, face width and length, and the distance from the front to the pritchel hole and the hardy

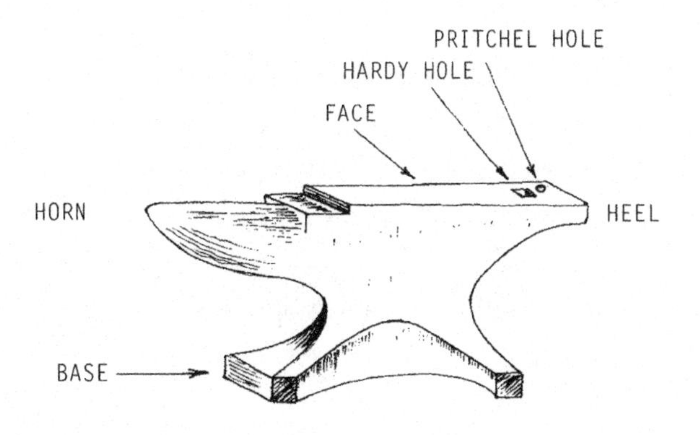

Figure 3.1 (facing). Forging a chain link.
Figure 3.2 (above). Use the anvil as a measuring device.

hole. Unless you are working to close specifications, measurement is a relative thing and not necessarily tied to feet or inches. Your anvil can then be a quick measuring reference.

- Steel expands when heated and shrinks when it cools. This can be used to your advantage in certain instances or may cause you grief if not taken into account! (This will be covered in more detail later.)

- For clothing, wear either wool or cotton. Stay away from nylon, rayon, and other synthetics. The synthetics will tend to melt and stick to your skin when hot. Cotton and wool will just smolder.

- Footwear should be good heavy leather boots—no tennis shoes or sandals.

- Pants should be worn outside the boots, *not* tucked in. If pants are tucked in, it is too easy for a piece of hot iron or slag to fall inside your boot, resulting in a nasty burn.

- A good leather or canvas apron is also recommended.

- Keep your main pieces of equipment close to the forge, like the anvil, hot cutter, trip-hammer, leveling table, and leg vice. If you have a trip-hammer, set it up so the on and off switch is by the forge, *not* on the hammer. You will be amazed at how much more efficient it is.

- If you do not have a reservoir on your forge, it is a good idea to keep a can—a 3-pound coffee can works well—of water close to the forge. It is good for cooling chisels and punches, and it serves as an emergency fire extinguisher if clothing catches on fire.

- Ear plugs should also be a consideration because of the loud noise when using the anvil, hammer, and other equipment.

- Deadening the ring of the anvil is a safety precaution to protect your hearing. To deaden the ring, take a piece of U-shaped steel, about 12 inches long, and place it in the pritchel hole.

- All anvil tools should be on the front or sides of the anvil block, never on the back of it. You will be barking your shins on anything on the back side of the anvil block. (See figure 3.4 on page 34.)

- Keep a wire brush handy. When the piece is pretty much finished and while it is still red hot, it is much easier to clean the oxidation scale off of it and will give a cleaner appearance to your finished work. If you have an odd-shaped piece that is hard to brush, have a larger box of loose, coarse sand handy. Stick the piece in the sand and move it around several times. This will knock the scale off the hard-to-reach places.

- A center punch or chisel is a handy tool to have when marking a spot to be bent, twisted,

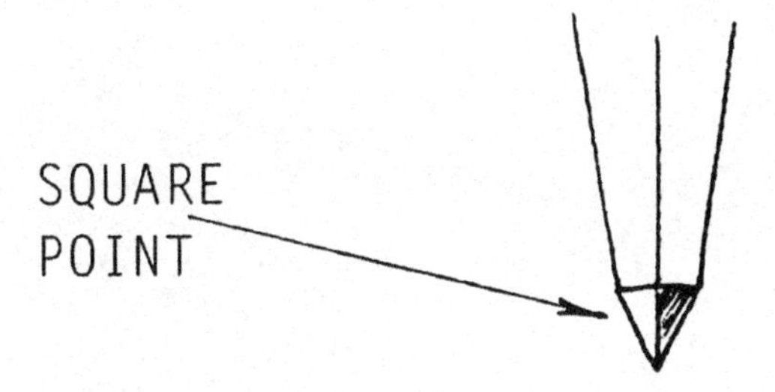

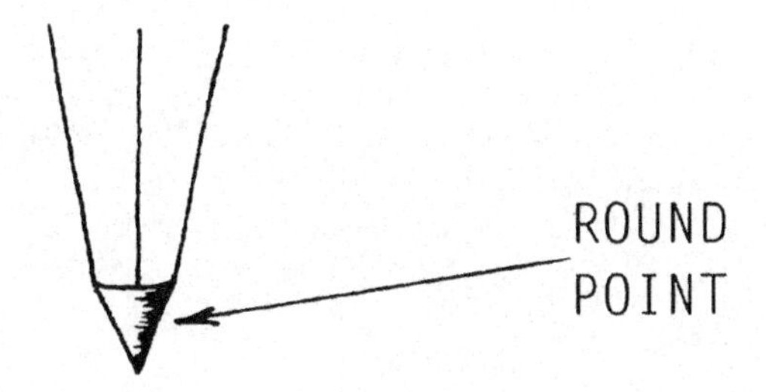

Figure 3.3. Tips on a center punch.

or split by heating. (Figure 3.3.) The mark will be easier to see at high heat. It also helps if the point is ground to a square. When marking a piece to bend with a chisel or center punch, mark it on the *inside* of the bend.

- Square stock will support more weight than round stock.

- Keep a notebook handy to jot down measurements, sketches, and other pertinent information. If you make something once, you will invariably have to make another one again at some point, so why reinvent the wheel each time? My notebook is one of the most used tools in my shop. It is also a great place to sketch ideas for future projects.

- *Put it back!* This cannot be stressed enough. In a true blacksmith shop, you will notice every tool has its own place, whether in a rack, on a table, or hanging on the wall. There is nothing more frustrating than getting into a project, pulling a piece of red-hot iron from the fire, and reaching for a tool that is not there. For that reason, no one works in my shop but me. Billy taught me that one of the first things he learned as an apprentice in his father's shop was to *put every tool back* after it was used.

- Keep your shop clean. This does not mean it needs to stand up to a general inspection, but keep the floor area around the forge, anvil, and trip-hammer clear. I have been in some shops that scare me. There is so much junk lying around, I'd be afraid to move from point A to point B for fear of tripping over something and hurting myself. One look at a shop and its layout will tell you a lot about the professionalism of the smith.

- One of the most useful pieces of equipment in the shop is a heavy-duty bench or pedestal grinder with 8- to 14-inch-diameter grinding

wheels. I am very protective of this tool in my shop. No one but me uses it because someone not familiar with using a grinder may inadvertently "catch" a piece they are grinding. What this can do is set up internal stresses in the wheel that can cause injury the next time you turn it on. I once had a 14 x 2-inch wheel explode on me for that very reason. Not fun. When using a large grinder to grind bevels or an edge on something, plant your feet and *do not move them* while grinding. You will be amazed at how even you can get a good bevel or even grind marks by just keeping your feet in the same place.

- When setting up your shop, the floor is best left as dirt or sand. It is much easier on your feet and back than concrete and a heck of a lot cheaper. In my shop in South Dakota, I did have a 5 x 8 foot slab of concrete on one side of the shop floor for straightening plow beams, hay bucker teeth, and the like. It also works well for laying out grates and grills where you need a large, level area to work.

- I would like to pass on one piece of advice that my mentor and master blacksmith, Billy Vogelman, gave me the first day I walked into the shop, *"Think twice and work once."*

Figure 3.4. Anvil tools should be on the front or sides of the anvil block.

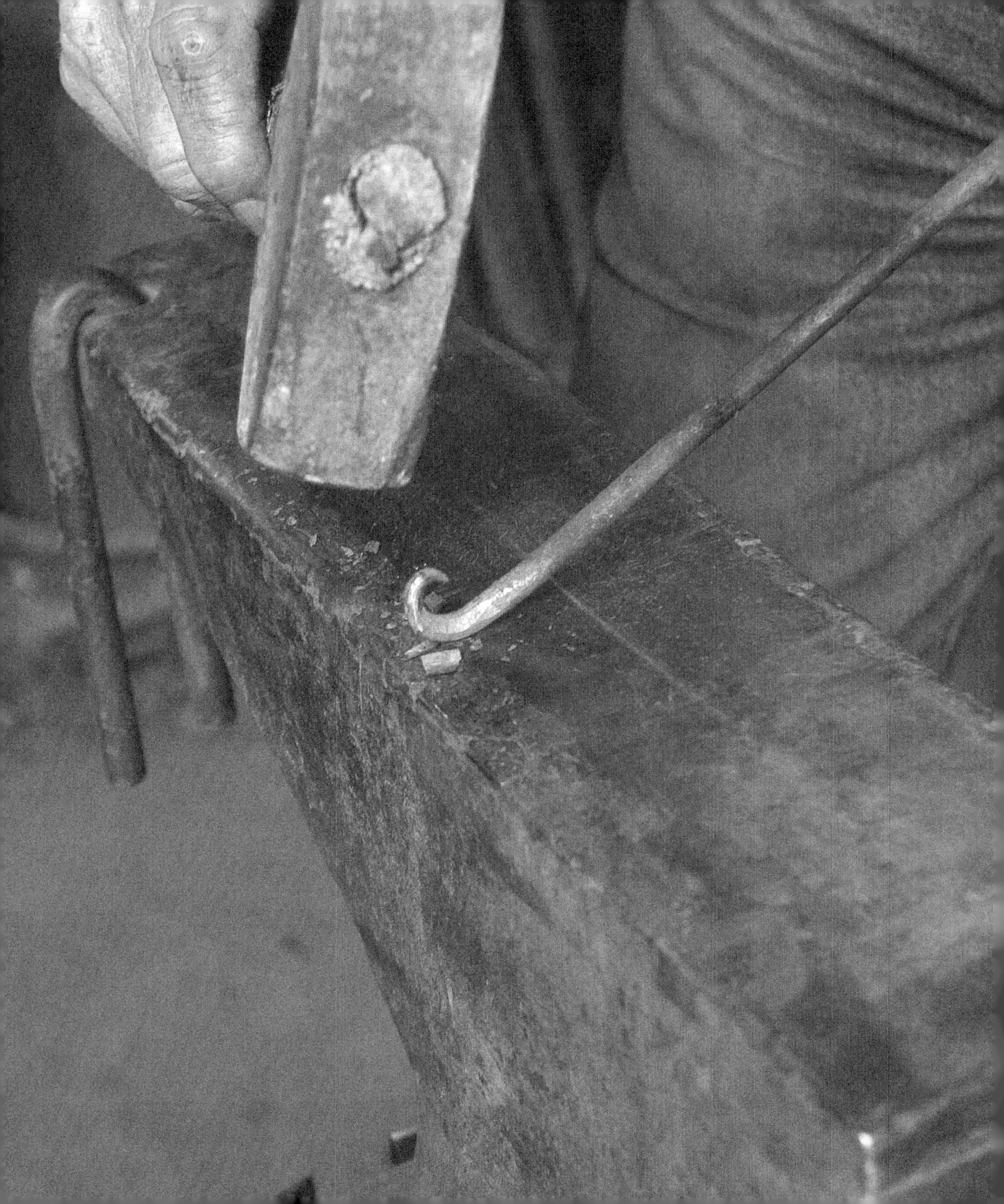

4 Basic Techniques

Before starting even the simplest of projects, there are certain basic techniques with which you should become familiar. You will see that the more advanced projects are merely basic techniques used in combination. As with all basic skills, they can be mastered only through practice. It is like learning a language; first you learn the alphabet, then simple words, and finally word combinations to form sentences, and so on.

The following are some of the terms we will be using during the course of the book.

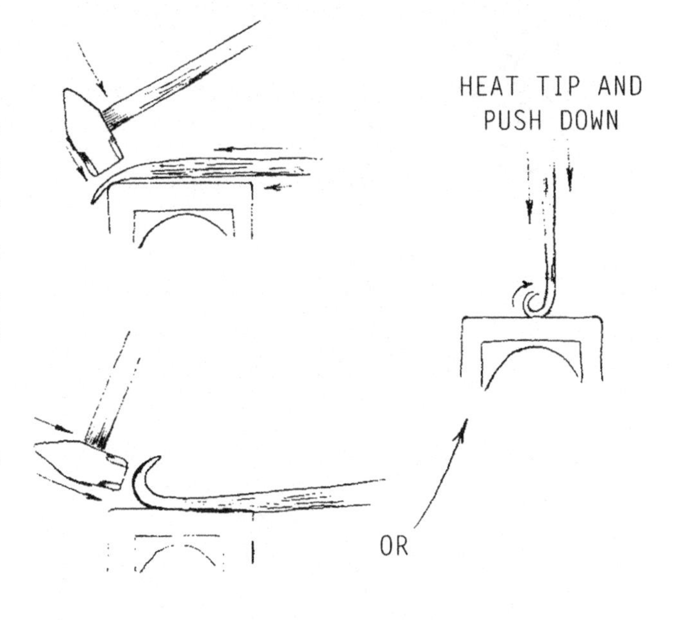

Figure 4.1 (facing). Curling.

Figure 4.2 (above). Two methods for curling.

CURLING

Curling is a decorative touch normally done after the end has been flattened or drawn to a point. (See Figures 4.1 and 4.2.)

DRAWING OUT

This is one of the simplest of the basic techniques, but it is used time and again from the basic projects to the complex. When working red-hot iron, keep in mind that it is like putty. Using the putty analogy, by squeezing and pushing the putty between your thumb and forefinger it gets longer and thinner

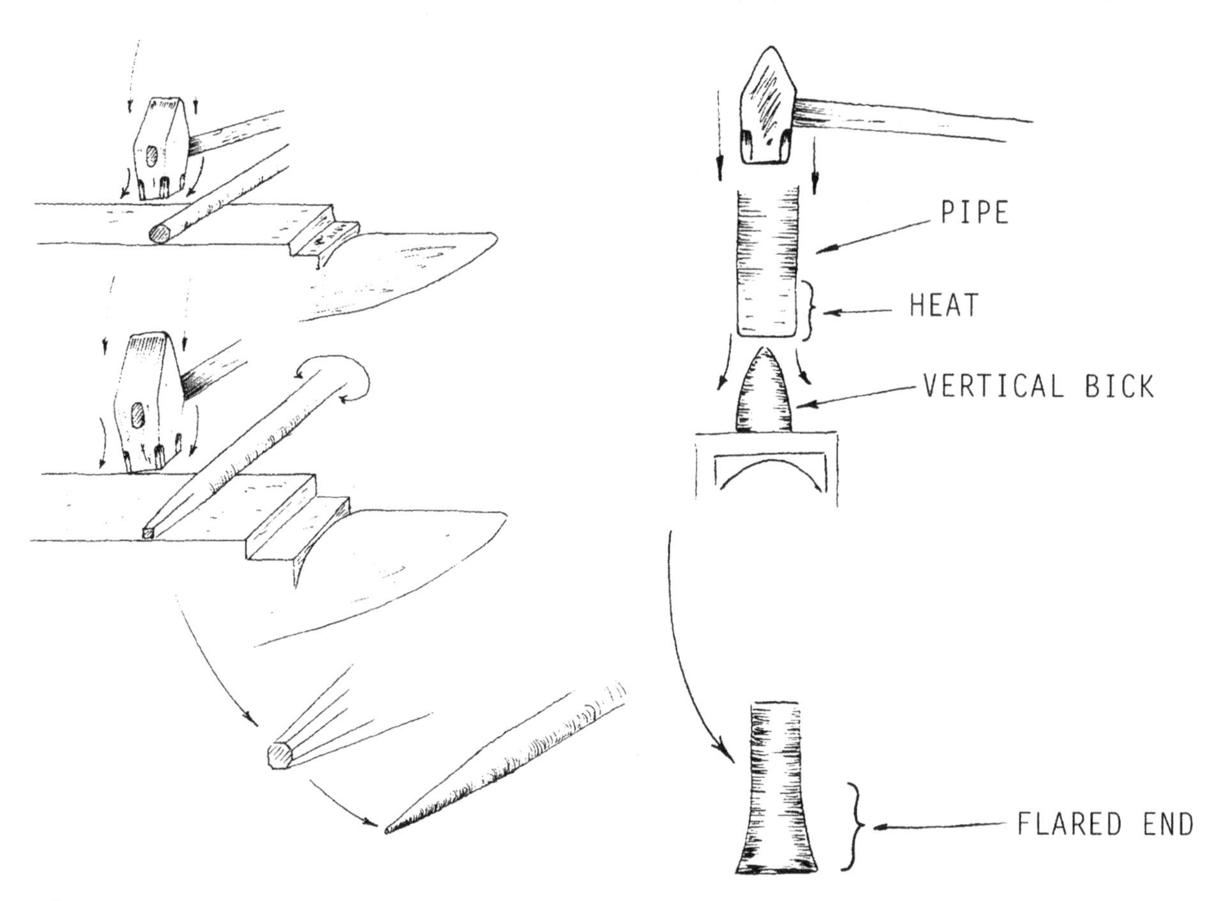

Figure 4.3. Drawing out.

Figure 4.4. Flaring.

(drawn out), the same as the hot iron. Push it together and it gets thicker (upset). When drawing out a rod that is round, it will always begin square. From the square it is worked to hexagonal, and then to the finished round configuration. When drawing the piece down, use the hammer in somewhat of a pushing motion as shown in Figure 4.3. This will facilitate the movement of the material toward the tip or end of the piece on which you are working. This is used in making everything from nails to axes and knives and is the process of tapering the end of the rod or piece being worked. Regardless of whether the piece is square or round, the tapering begins as a square.

FLARING

This method is usually used in reference to tubing or pipe where the edge is pushed out to be wider than the body, or flared. (Figure 4.4.)

FULLERING

In these methods, a fuller is used to spread, notch, or draw out hot iron. (Figure 4.5.)

HEAT COLORS

As the metal heats up, you will notice a change in colors from a dull red to bright red to orange red to yellow red to yellow and finally to white heat, which is the fine line between forge welding heat and burning the metal.

PUNCHING

This process may be used instead of drilling on all but the smallest pieces. Remember to move quickly, before the heat is lost. On larger pieces such as hammers, it is important to keep the tip of the punch cooled off to avoid distortion. Drop a pinch of cinders in the hole once it is started to keep the tip of the punch from getting too hot. The advantage of punching is that any type of hole may be punched: round, square, oval, or diamond. (Figure 4.6.)

RIVETING

This is a method of joining two pieces of metal by using pins hammered flat on each end. (See Figure 4.7 on page 40.)

SCARF OR SCARFING

This refers to the process of thinning an area down prior to welding two pieces together. (See Figure 4.8 on page 40.) It makes for a smoother appearance on the finished piece after the weld is done.

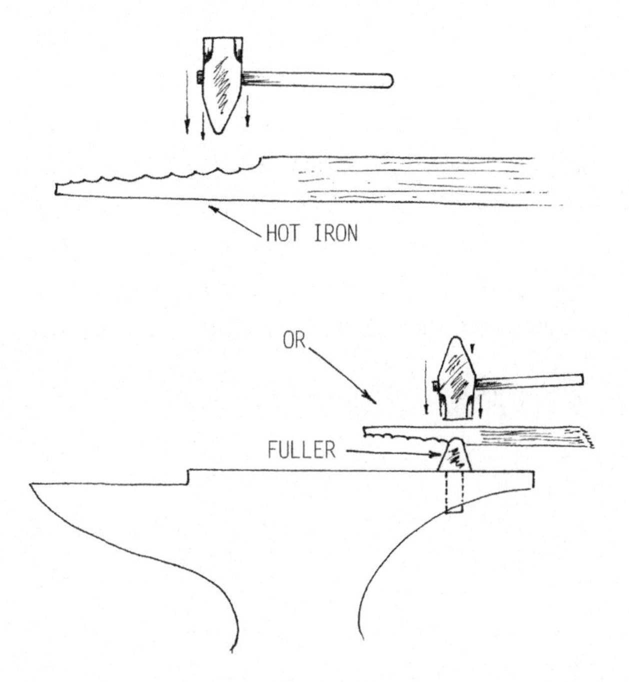

Figure 4.5. Two methods for fullering.

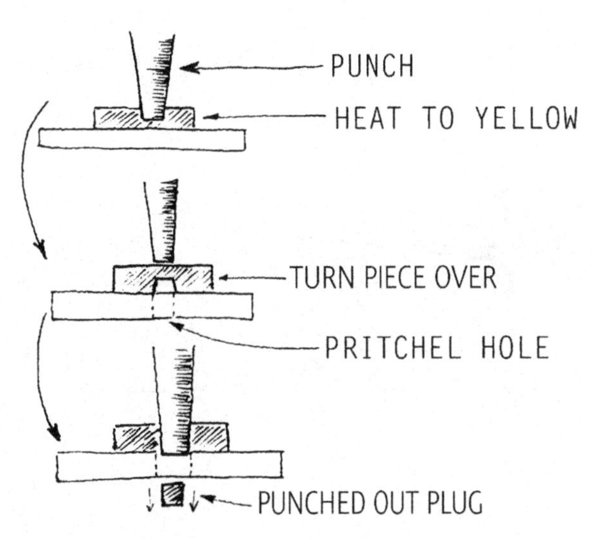

Figure 4.6. Punching.

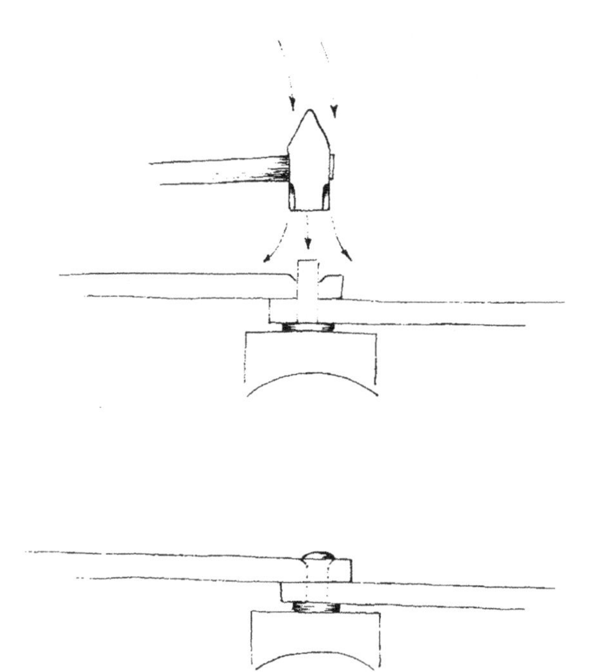

Figure 4.7. Riveting.

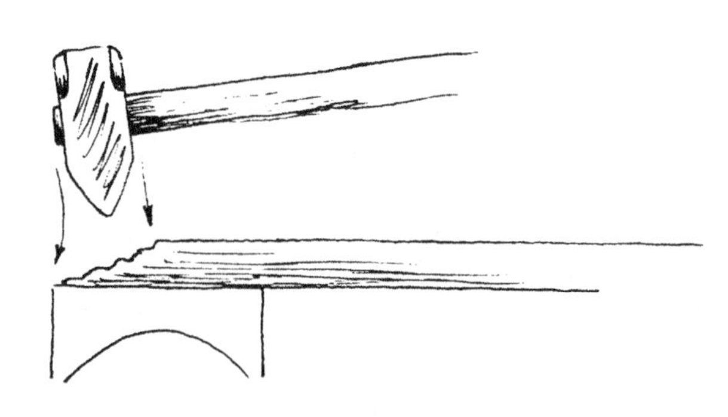

Figure 4.8. Scarfing.

SPLITTING

This process has numerous applications, from decorative work to toolmaking. Next to drawing out, this is probably one of the most used procedures. The important thing to remember here is to keep the cutting edge of the chisel cool. Being relatively thin, it will tend to get hot and become distorted very quickly. This is especially true when cutting or splitting heavy pieces, such as axes. The splitting must be done from both sides. On lightweight material, however, splitting may be done from one side only. (See Figure 4.9.)

TWISTING

Twisting adds a decorative as well as utilitarian touch to various projects. (Figure 4.11.) It is often used in ornamental work such as andirons, grates, railings, cooking utensils, or tools that have a handle. Keep in mind that the length and tightness of a twist is determined not only by the number of turns, but also by the length. A long heat will give a long twist. Twisting can only be done on square, hexagonal, or rectangular stock. It cannot be done on round stock.

UPSETTING

V

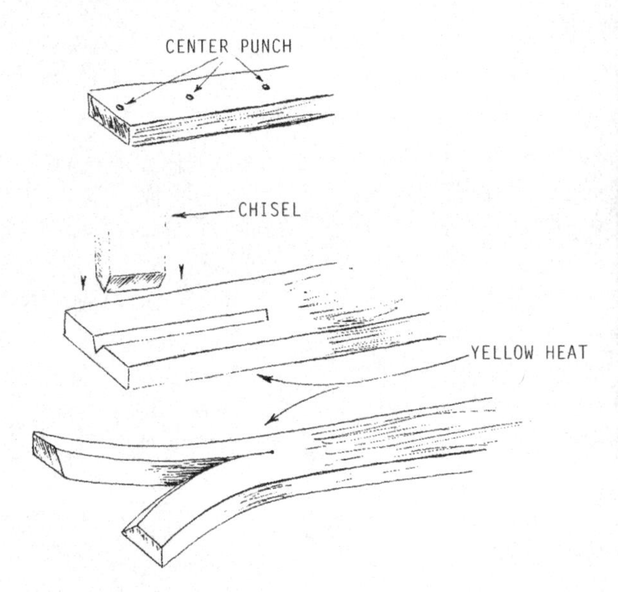

Figure 4.9. Splitting.

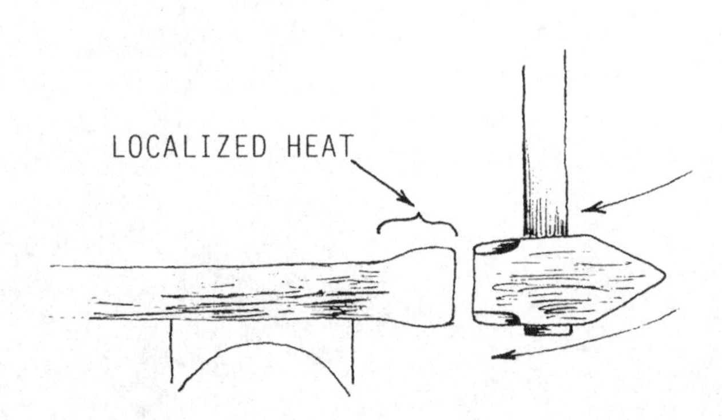

Figure 4.10. Upsetting.

Figure 4.11. Twisting.

5 Firing Up the Forge

Now that the basic techniques have been covered, it's time to get the forge fired up. The fire is the single most important tool used in blacksmithing. As my mentor, Billy, used to tell me, "The fire does the work, you just make sure it's done right." In the old days, sawdust, wood chips, or kindling of any sort was used to start the fire. There are many ways it can be done. Some that I have seen defy description. Perhaps some folks feel making a ritual of starting the fire, which is fine up to a point, enhances the mystery of blacksmithing.

We will assume that the fire you are starting is the very first one, which means you will be using "green" coal, or coal that is not coked. Let me explain a bit. Once the fire is started, you will notice that there is a large amount of greenish-gray smoke and a dark red flame. This is the green coal burning. It gives off almost no heat and is impossible to use for forge welding. As the fire continues to burn, you will notice that it begins to burn a very bright yellow-white in the center and that very little smoke is being produced. Your fire is now at working heat.

Figure 5.1 (facing). Starting the fire using crumpled newspaper.

Back to starting the fire: the easiest way is to ball up three or four sheets of newspaper, tear up some cardboard to lay on top of the newspaper, light the paper, push some, but not all, of the coal on top into the fire box, also known as the "duck's nest" or tuyere, and start the blower. A word of caution here: Don't get the air blast going too strong or you'll blow the paper right out of the fire pot! Take it slow and easy. Once the fire is going, be sure to pile coal around the outside of the fire so you have a ready supply to work toward the center. Never shovel green, uncoked coal directly on top of or into a fire that is at working heat because all it will do is give lots of flame and smoke with no heat.

A couple of shovelfuls of green coal should suffice. Once the fire is going, you will find that you have to periodically knock the coke into the center of the fire with a hooked poker. This will come naturally as you become proficient in handling the fire. Be patient. All of this will take practice and will eventually become second nature. Working the fire properly is extremely important. After the fire is out, you will have coke left over that will start much easier than green coal, or coal that has not coked up. That is what you want to use for starting

Figure 5.2. Coke.

your next fire. Keep in mind, the coal will coke up around the edge of the fire—this is what you want to burn, *not* the green coal.

Surrounding the "blast furnace" center of your fire, you will notice clumps of almost pumice-like coal. This is the coke or coking that was referred to earlier. It is basically a clump of pure energy, having had the impurities burned out of it. (Figure 5.2) Remember the smoke and red flame?

This is why the fire is always worked toward the center, continually pushing the coke, which forms around the edge of the fire, toward the middle.

Perhaps I should mention clinkers here. Clinkers are the impurities and residue left over after the coal and coke have been consumed. They will have the appearance of hard, glasslike clumps when the fire cools. (See Figure 5.3.) In a working fire, they will show up as dark gray or black "dead spots" down in the center of the fire. This is what is referred to as a dirty fire. Before any type of welding can be done, these clinkers must be cleaned out of the fire. I have found it easier to let the fire sit idle (no air blast) for a few minutes, and then, with a straight poker or paddle, lift the clinkers out and throw them off to the side.

Generally, forge coal is worked damp by adding water from the slack tank (a 5-gallon water bucket or tub for quenching hot metal). The consistency of oatmeal works well for me. You may want to work it more or less wet. By working the coal wet, two things are accomplished: (1) the coal tends to coke up a little better, and (2) you can shape the fire more easily. Shaping a fire is arranging it according to its purpose. For example, use a long fire for knife blades and a deep fire for welding.

Remember, the fire does the work; you merely guide it. Now that the fire is going, it is time to get started.

Figure 5.3 (facing). Clinkers.

6 Forge Tools

The projects we will cover here range from the simple to the more complex and will require a minimum of tools. We will start by making some of the tools you'll find useful for tending the forge fire and move on to making tongs for holding the hot metal at the forge.

HANDLES

Stock: 3/8 x 30 inch, square

The first thing in making fire tools is to make the handle. Three simple yet effective handle styles are round, teardrop, and twist. (See Figure 6.1.)

The round handle is made by taking a 3/8-inch rod, about 30 inches long, and heating approximately 4 to 5 inches of one end in the fire. Bend the 4 or 5 inches over the edge of the anvil at a 40-degree angle. Reheat. Now bend this section around the horn of your anvil in the opposite direction to form a closed circle.

The teardrop is made in basically the same manner but in the shape of a teardrop instead of round.

Figure 6.1 (facing). Types of forge handles: teardrop (top), round, and twist.

The twist handle is just what its name implies: twisted. Merely heat a 3- to 4-inch section of the handle to a good red heat, put one end in the vise, and use an adjustable wrench at the other end to twist the handle.

HOOKED POKERS

Stock: 3/8 x 30 inch, square

Using a piece of 3/8-inch stock, make the desired handle. Draw out the other end to a short point and make a right-angle bend about 2 to 3 inches back from the end by bending it over the edge of the anvil. (See Figure 6.2 on page 48.)

STRAIGHT POKERS

Stock: 3/8 x 30 inch, square

This is a straight piece of stock the same dimensions as the hooked poker. Point one end and leave it at that. This poker is used primarily for sliding down the edge of the fire and prying up to "open" the fire.

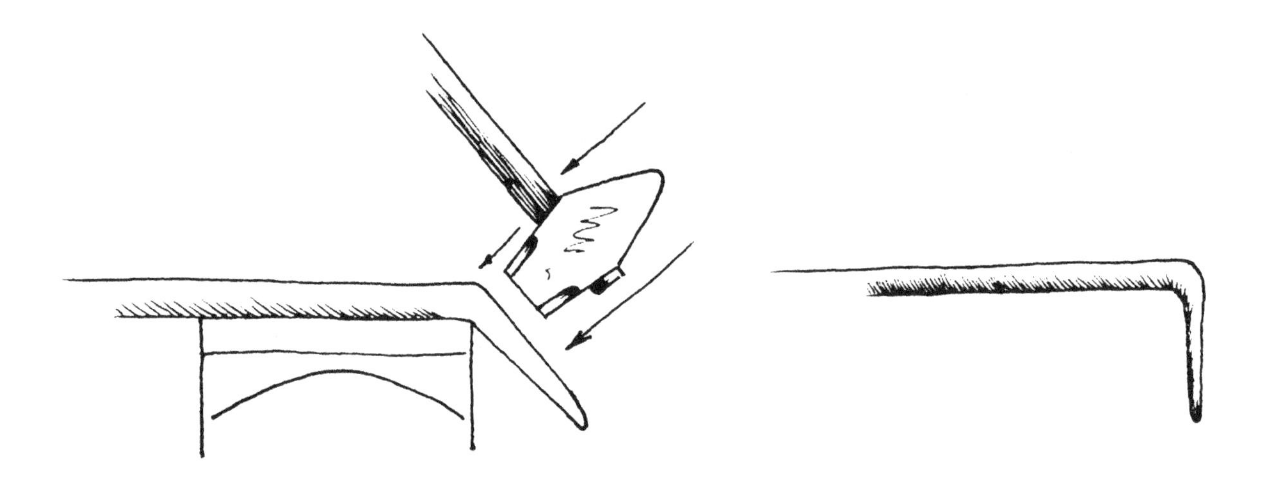

Figure 6.2. Hooked poker.

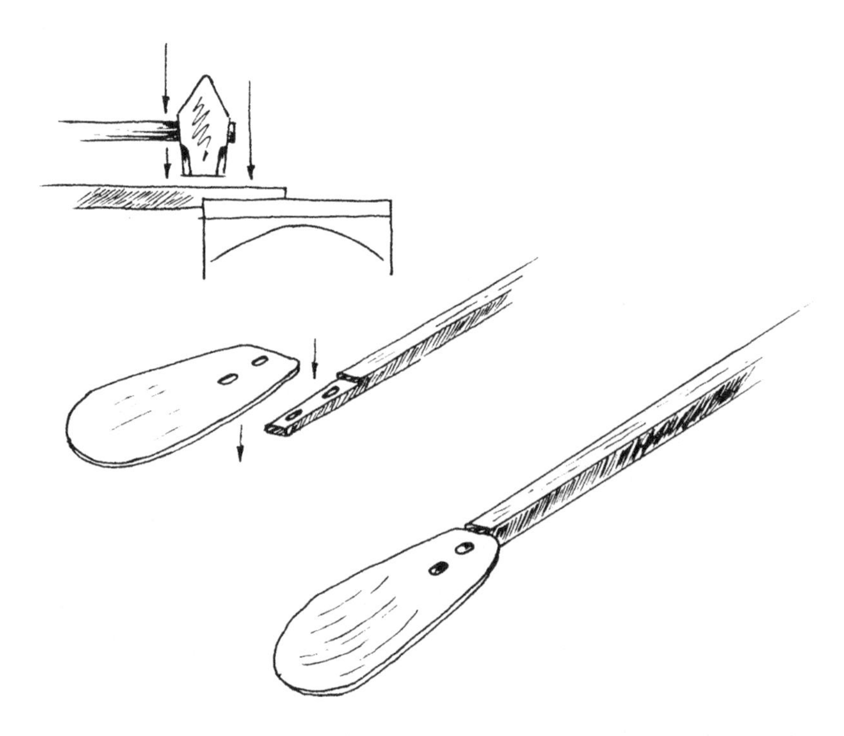

Figure 6.3. Shovel or paddle.

SHOVELS

Stock: 3/8 x 30 inch, square

This is more like a small paddle. It is used primarily for shaping the fire. Make the handle first. Flatten the working end of the stock and drill two small holes the size of the rivets you plan to use. Take a piece of 8-gauge sheet metal and shape the shovel to the desired size and shape. Drill two holes in the sheet metal that correspond to the holes in the handle and rivet the two pieces together. (Figure 6.3.)

TONGS

Stock: 2 pieces 1/2 x 18 inch, square

Tongs are as varied as the projects in which they are used. They may be reshaped to fit a particular job. Tongs may be found at farm auctions and flea markets, or bought new from suppliers around the country. If you can't find used ones and don't want to buy new ones, tongs can be made to fit your needs. It is best to buy several general-use tongs at first. As you become more proficient on the anvil, you can make whatever type you may need.

The pair of tongs described here is a very basic or general pair.

Tongs are really nothing more than a pair of extra large, elongated pliers. They are an extension of the hand and may be used to hold, move, or shape the piece of material being worked in the forge.

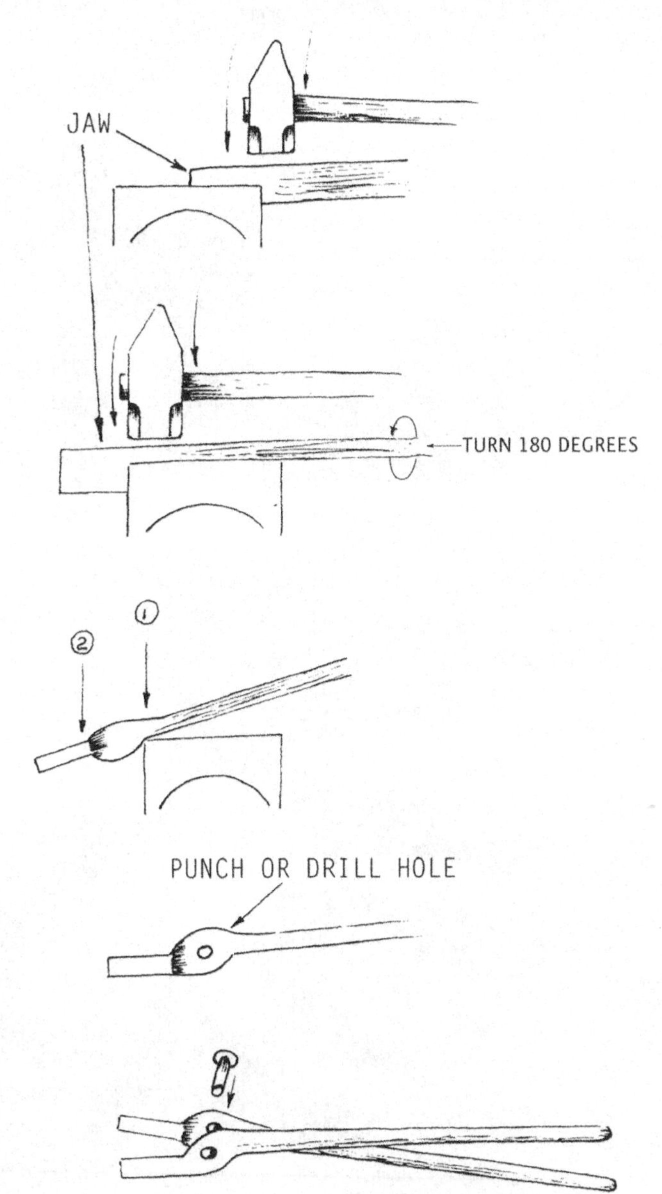

Figure 6.4. Basic tong construction and assembly.

Although there are literally hundreds of designs, they are all made in the same basic manner. Tongs found at Roman archeological sites are identical to those in use today.

There are several things to keep in mind about tongs before we begin: (1) they must pivot easily without being sloppy; (2) the jaws should not be too thin; and (3) use only *mild steel* for their construction. After the tongs have become hot and are cooled during use, high-carbon steel may become brittle and break.

Begin by heating 1 to 1 1/2 inches of one end of the stock to a good yellow heat and flatten to approximately 3/8 inch. This will be the jaw portion of the tongs. Now, reheat just behind the jaws, turn the stock halfway over, or 180 degrees, and flatten a quarter-size area approximately 3/16 to 1/4 inch thick. This will be the area where the hole will be drilled or punched to join the two pieces.

Remember, you are making two right-hand pieces. Once this is done, offset the handle over the edge of the anvil as shown in Figure 6.4 on page 49, and draw out the handles to the desired thickness. Many of the old tongs had the handle made of two pieces forge welded together.

Place the two pieces together and rivet them. For final adjustments to the jaws and handle alignment, reheat and adjust accordingly. To enhance a good fit at the pivot point, rivet, heat to a cherry red, move the handles so they work freely, and, as you cool them in the slack tank, keep moving them until cool. You now have a pair of smooth-working tongs that are not sloppy.

This is the same procedure, using smaller stock, of course, for making pliers and scissors.

Figure 6.5 (facing). Various types of tongs.

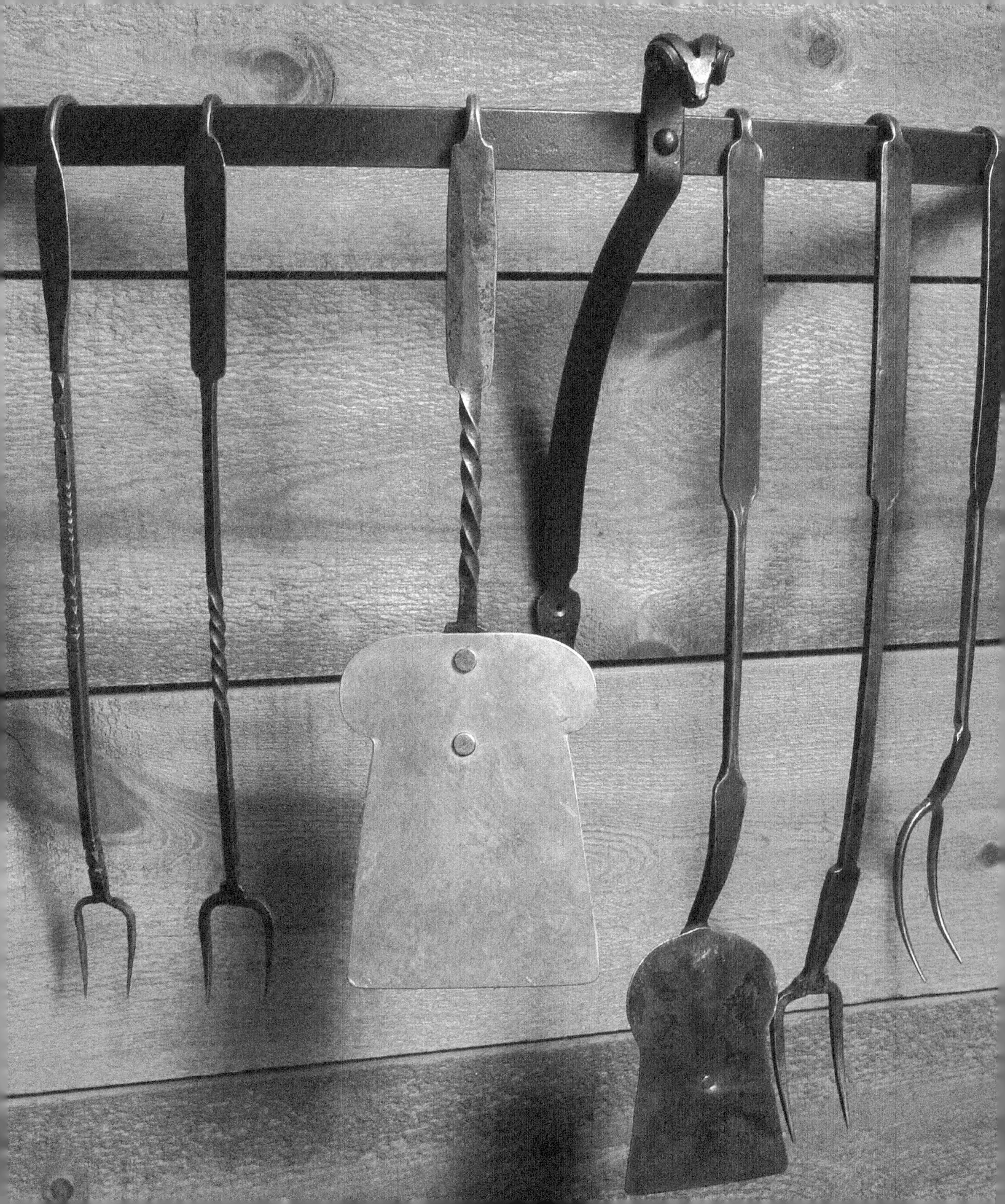

7 Basic Projects: Kitchen and Cooking Gear

While getting settled on your homestead or just out camping, you need to eat. To this end, you need some cooking gear.

HOOKS

Pot Hooks or S Hooks

Stock: 1/4 inch x 2 feet, round or square (mild steel)

Pot hooks or S hooks, as they are sometimes called, have been in use for thousands of years and are one of the most used (and most often lost) pieces of hardware. They may be made of round or square stock, and may be fancy or plain, depending on your preference.

To make a basic, plain pot hook, take the piece of 2-foot-long stock. (The reason for the extra length added onto the stock for all projects here is so that you have parent stock. Parent stock is the extra-length of the stock that is cut from the finished piece. Do as much of the work while the piece is attached to the parent stock. Tongs are good but won't replace being able to hold onto the piece with your hand while working it.) Heat about 3 inches of the end to a good orange-red heat. Take

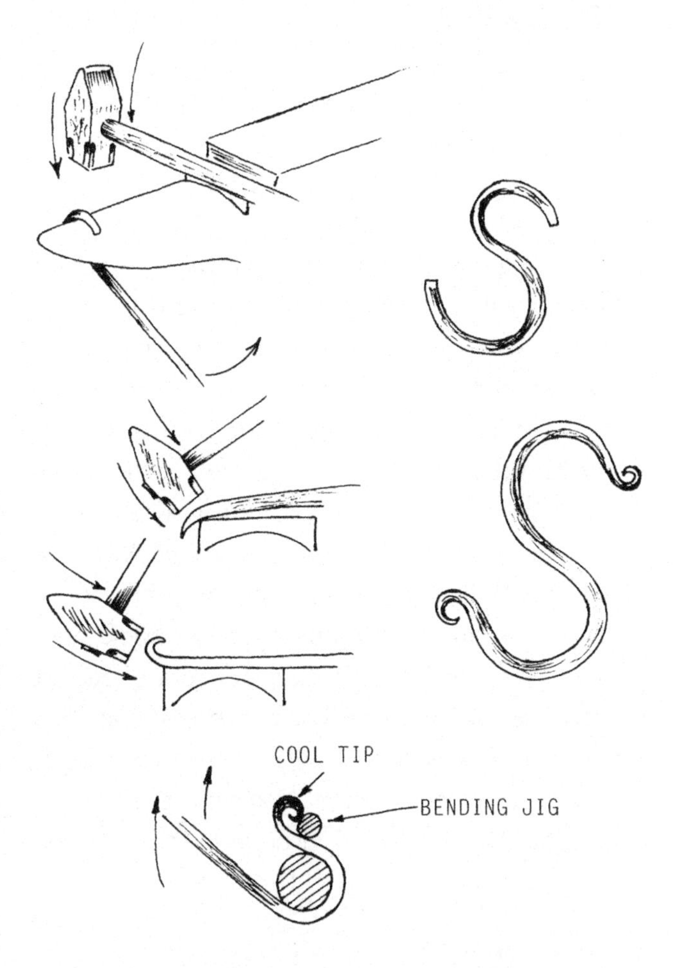

Figure 7.1 (facing). Kitchen utensils.
Figure 7.2 (above). Pot hooks.

the heated end and bend it *over* the horn of your anvil as shown in Figure 7.2 on page 53. From the bend, measure off approximately 6 inches of the stock and cut it off on the hardy. Holding the half-completed pot hook in your tongs, heat up about 3 inches of the straight end and bend it over the anvil horn in the opposite direction from the first bend. You now have an S-shaped hook—very plain yet very serviceable.

Several things may be done to improve the hook's aesthetic quality. Before making the first part of the hook, which is the large bend, draw out the end to a nice taper. (See Figure 7.2 on page 53.) Now, heat just the tip and curl it, and then reheat approximately 3 inches of the end, dip the small curl in water to cool it, and bend the first part of the hook as shown. Repeat the process on the other end. Always remember that the small curl for the tip will be bent in the opposite direction from the larger bend for the hook. If using square stock, you can also put a twist in the middle between the two hooks. Follow the same procedure as above to put the small curls on the tip of the hook. After that, heat the middle of the piece to a good red, lock one end in the vise, and give it at least one full twist. Before it cools, make sure everything is straight, and then cool it off. You now have a simple but graceful item. As with all projects, the more you practice, the more proficient you become.

CAMP SETS

Stock: $^1/_2$ x 40 inch, square ($^3/_8$-inch stock may be used for a lighter-weight set)

The basic camp set or cooking irons, as they are sometimes called, is the epitome of utilitarian simplicity. It is composed of three parts: two uprights and a crossbar. A small trammel hook and/or S hook may also be used with the camp set.

Begin by heating one end of each of two of the pieces and either draw out to a point and curl or merely flatten and curl. (Figure 7.3.) Measure back approximately 4 inches and mark the *inside* of where the bend will be. This will be the side facing the front of the curl when you bend it back on itself. Heat the area at the mark, and using either the edge of the anvil or a scroll fork, start the bend to 90 degrees or more, and hammer this section back on itself for the first 1 $^1/_2$ to 2 inches. Reheat the 2-inch portion with the curl on it, being careful not to burn the curl, cool the first 1 $^1/_2$ to 2 inches on the end, and then bend the heated piece up in a hook-like projection. You may close it completely or leave it open, depending on your preference.

You can see the area where the two pieces that were bent back on each other can be pounded on without distortion. These two pieces then will be uprights. The next piece we will make is the crossbar. As with the straight poker for the forge tools, make a round ring on one end proportionate to the length of the rod. I've found a 1 $^1/_2$- to 2-inch-diameter ring works well. For a 2-inch-diameter ring, use 6 $^1/_2$ inches of stock. By making a ring on one end, you will, by hooking the two uprights into it, be able to make a tripod configuration to cook with, which will require a trammel hook or S

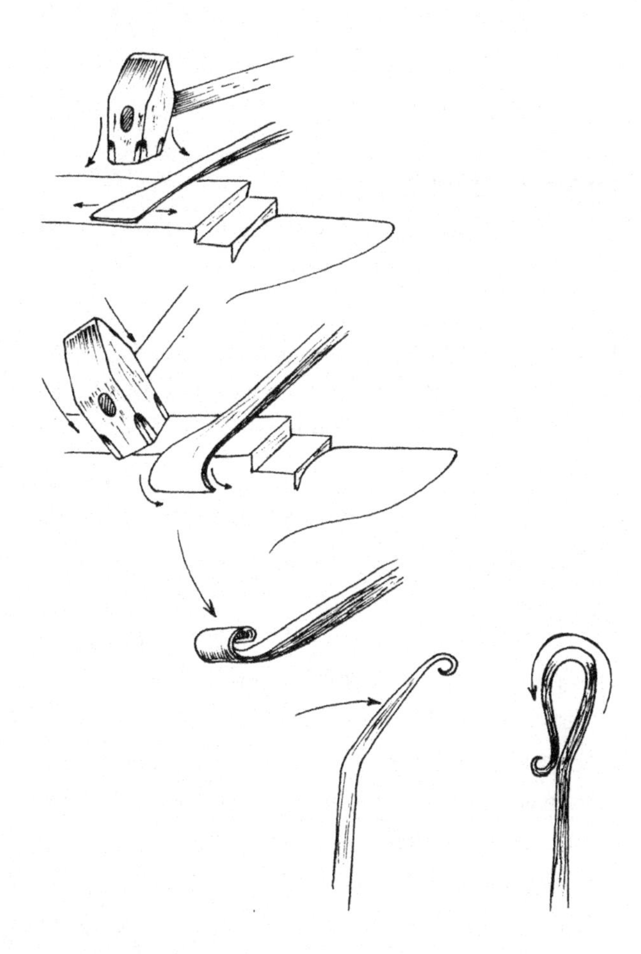

Figure 7.3. Finial for the camp set.

hook to hold the pot or kettle being used. All that remains to be done is to draw out a rather blunt point on the end of each of the three pieces.

If you want to add a decorative touch, you can heat a 2- or 3-inch section in the middle of the uprights and twist it. Make sure the twist is in the same place on each one for a uniform appearance. Do *not* put a twist in the crossbar as this will weaken it.

COOKING UTENSILS

Forks

Stock: 3/8 x 14 inch, square (mild steel)

Cooking forks come in various lengths, but normally a length of 12 to 14 inches works well for cooking at the hearth or campfire. I would recommend using 3/8-inch-square stock, as it will be easier to work with.

Once again, begin by making the handle, which in this case will be a simple twist and hook. Draw the end out to a point and curl as with the S hook, and then bend the opposite way (after cooling the curl). This small hook serves a twofold purpose: (1) it is used to lift pots off the fire, as well as pot lids, and (2) it is also used to hang up the fork. From here, decide what length you want the fork to be and cut it off. Keep in mind that as you draw out the tines of the fork it will increase the length of the fork slightly. To allow for this, I usually cut the fork stock 1 to 1 1/2 inches shorter than the desired length.

When you have determined the length, you can proceed in several ways. Draw the end out to a long point and then flatten slightly. After the piece is heated, hold it in place on the anvil with the holdfast and split the end for about 2 to 3 inches with a thin chisel.

Use a soft block under the work to protect the anvil face. This splitting may take several heats, so watch that you don't burn the tips off the fork when reheating in the forge.

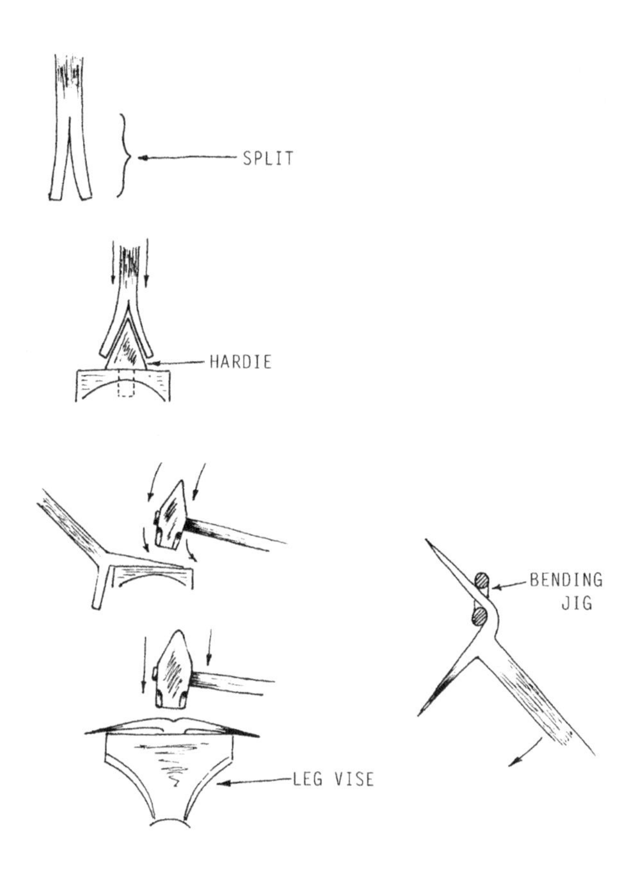

Figure 7.4. Method for forming fork tines.

Once the piece is split, open the split by pushing it down over the edge of the anvil or hardy. (See Figure 7.4.) You can now heat the tines and bend them to the shape you desire for the fork. Remember to bend them so the ends of the tines come out even. They may be bent with a square or round shoulder.

The other method of construction is to flatten the end for the tines, split it, and then draw out each individual tine of the fork to the desired shape and length. Try both methods and see which one works best for you.

Spoons/Spatulas

Stock: 3/8 x 16 to 18 inch, square or round (mild steel) and 1/16-inch or 15-gauge sheet metal

These two cooking utensils are grouped together because their method of construction is identical; only the shape is different. Using this basic procedure, several varieties of serving spoons may be made, such as a ladle and the slotted or strainer spoon.

To start, flatten one end, of the stock and bend that part in an open curl. Come back about 4 inches from the end, heat and bend in the opposite direction until the curl touches the shank. You now have a very plain, serviceable handle. After this is done, hold the piece by the handle, heat up the other end, and flatten. After this is cooled, drill two holes for rivets (3/32 to 3/16 inch works well) and countersink them on the back side of the ladle. This will give a good, tight fit for the rivets. (See Figure 7.5.)

For the spoon or ladle, cut a piece of 1/16-inch or 15-gauge sheet metal in the shape desired for the spoon, heat it in the forge to a good cherry red (be careful not to overheat and burn it), and pound the spoon into shape with a ball-peen hammer. This may be done in several different ways. A depression can be carved in the end of a large stump and

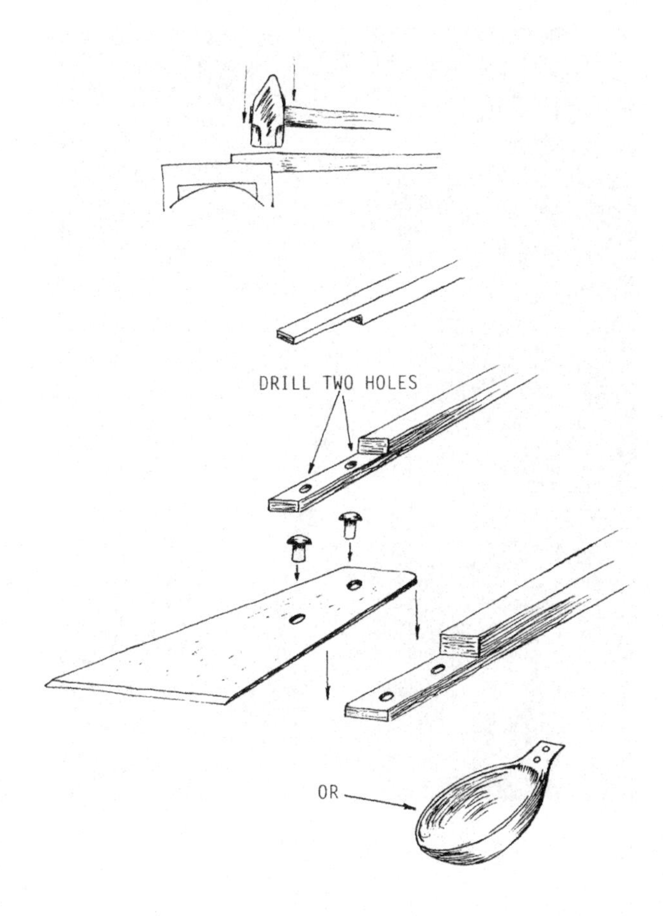

Figure 7.5. Attaching the handle to a spatula or ladle.

used as a forming die, or you can take the piece, and after it is heated, put it on the ground and shape it. If you are lucky enough to have a swage block, that works perfectly.

After the spoon is shaped, hold it on the handle to line it up, mark the holes for the rivets, and drill them. All that remains to be done now is setting the rivets, and the spoon or ladle is finished.

For a spatula, you merely cut the shape desired for the flipper, mark the rivet holes, drill, and set the rivets.

Another method that will work, although it takes more practice, is to forge weld a wider piece on the handle, say 1/4 x 1 1/2 x 4 inches flat. After it is welded in place, heat to a high yellow and flatten the piece out to 1/16-inch thickness. Cut and form to the desired shape for a spatula, spoon, or ladle. As I said, this process requires a bit more skill and practice than the riveting method, but it produces a very smooth and aesthetically pleasing piece. Finish out the fancier welded spoons and spatulas by filing and polishing.

For smaller spoons, you can make them all in one piece; 3/16 x 1-inch stock works well. First, draw out the handle portion to 4 inches or 5 x 1/4 inches either round or square. Cut from the parent stock, leaving 1 inch of stock to form the spoon itself. Flatten this to about 1/8-inch thickness, file or grind to the desired shape, and disk to the proper depth.

Skewers (Asadas)

Stock: 3/16 x 1/2 x 24 inch, flat (mild steel)

For lightweight, primitive camping, it's hard to beat a set of skewers. The skewers, combined with the S hooks, are virtually all that you need. They are most versatile if used in sets of three. In this way they may be used as individual skewers, a grill, a camp set, or a tripod. They have been in use worldwide for thousands of years. A nice, lightweight set can be made from 3/16 x 1/2 x 18-inch flat stock. Length may vary according to your personal taste, but 18 to 24 inches is a handy size.

To begin, heat 4 inches of the end and make a 45-degree bend as shown in Figure 7.7. The bend should be made at the point where the heat ends. Now, quickly, while you still have heat, bend the piece that still has heat back in the opposite direction. It should look like a ring on the end. Take a light heat and finish shaping. You may either close the end or leave a small opening. By leaving a small opening on one of the three skewers, you can hook them together and make a tripod.

Now that you have the handle portion completed, the next step is to put a twist just below the ring and then point the end. With the completion of several S hooks and a set of skewers, you have a basic, primitive camp set that will suffice for all but the most elaborate cooking methods.

TRAMMEL HOOKS

Stock: 3/16 x 1 x 24 inch, flat (mild steel) and 1/4 x 10 inch, round

The last bit of cooking equipment that we will cover is the trammel hook. It was most often used at home in the cabin in conjunction with the cooking crane in the fireplace. Many trammel hooks were quite elaborate and complicated depending on the wealth of the household. The one that we will describe here is of the simpler design.

Figure 7.6. Finished skewer.

The techniques used will be drawing out, curling, punching, and bending.

Heat one end of the 2-foot stock in the fire and draw it out to a long, even taper as shown in Figures 7.8 on page 60 and 7.10 on page 61. This will be the top of the trammel hook. Curl the very tip in a small curl and make the large curl or hook as shown in Figure 7.11 on page 61. Be sure to "offset" the hook so the direction of pull is centered on the hook and not to one side. (See Figure 7.12 on page 62.) This is for strength as well as aesthetics.

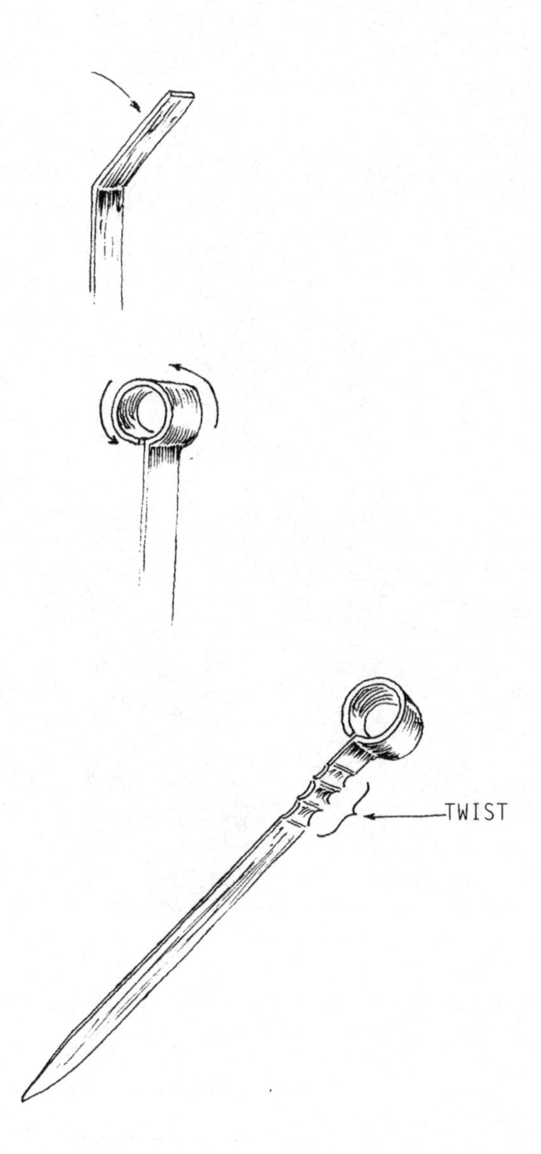

Figure 7.7. Steps in making the skewer handle.

From the shoulder below the hook, measure about 8 or 9 inches (longer if you want a longer hook) and cut it off. The next step is punching the holes. Decide how far apart you want the holes and mark them with a center punch. (See Figure 7.13 on page 63.) A word of caution here: do not get the bottom hole too close to the bend, or you will be unable to insert the long hook. Punch the holes with a 5/16- or 3/8-inch round punch. You will be using 1/4-inch stock for the long hook, but it's best to have the holes oversized so the "catch" hook will slide in and out of the trammel bar easily. The reason for this will be evident when you are standing in front of your fireplace or a smoky campfire holding a 10-pound pot of stew, which is probably scalding your knuckles, and you're fumbling around trying to get the catch hook back in one of the holes. If it doesn't happen to you personally, just let it happen to the cook once. I'm sure you will be advised in no uncertain terms how much larger the holes should be!

The very bottom hole should be made about 1/2 inch from the bottom. Now that all the holes are punched, heat the bottom inch or so (make sure it's above the bottom hole), lock it in the vise, and make a right-angle bend as shown in Figure 7.14 on page 63. Be sure to make this bend as close as possible to the hole without distorting it. This hole is now the lower guide, so to speak, for the long hook. The long hook is nothing more than a 10-inch hook with a right-angle bend on one end. When making the right-angle bend, it is a good idea to bring it a little past 90 degrees, as shown.

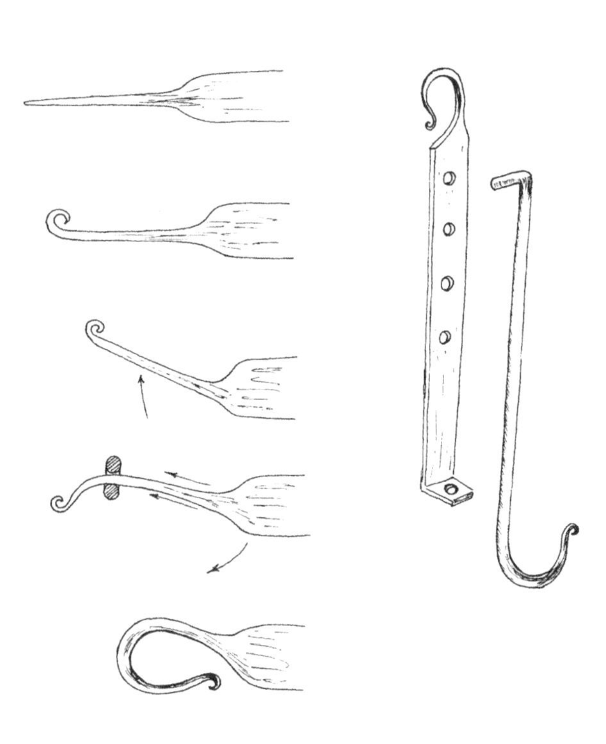

Figure 7.8 (above). Trammel hook assembly.
Figure 7.9 (right). Completed trammel hook.

This will keep the hook from slipping out. Needless to say, this could be rather embarrassing. When you are making the hook, be sure to have it facing the opposite direction from the upper hook of the trammel bar. As mentioned earlier, there are many designs for trammel hooks, and this is one of the simpler ones. Don't be afraid to try some of the more complex pieces.

Figure 7.10 (top). Drawing out the top hook of the trammel hook.
Figure 7.11 (above). Bending the offset for the hook.

Figure 7.12. Bend the top hook using a bending jig.

Figure 7.13. Punch holes in the body of the trammel hook.

Figure 7.14. The completed bend at the bottom of the trammel hook.

8 Basic Projects: Amenities and Tools

BEAM HOOKS

Stock: 1/4 inch x 2 feet, round or square (mild steel)

Another handy yet simple item to make is the beam hook. (Figure 8.1.) This is nothing more than a modified S hook. (See page 53.)

Begin by making the first hook and then, instead of making the other hook, merely draw the stock out to a good square point. Now, come back several inches and make a right angle bend. (Figure 8.2) Two things to keep in mind here: it is best to go a little more than 90 degrees on the bend, as it will tend to "lock" the hook into the wood. The other thing is to make sure you have enough room *above* the tip of the hook to pound it into the wall or beam.

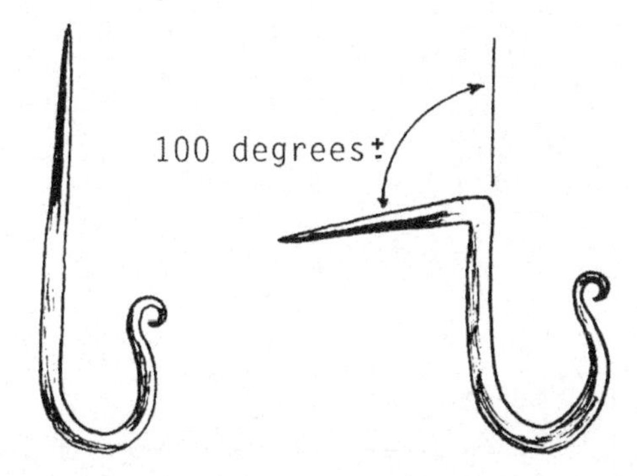

Figure 8.1 (facing). Beam hook, left, and large S hook.
Figure 8.2 (above). Beam hooks.

Figure 8.3. Hand-forged nails.

NAILS

One of the most used items on the homestead is the common nail: a basic, no frills, utilitarian accoutrement. We take them for granted today, but in years past, they were almost a luxury. Old buildings were burned down to retrieve the nails so they could be reused. As a sign of wealth, designs made out of nails were incorporated on the front doors of some of the early homes. It showed the owner could afford to use nails for something other than building. In the early days in Scotland, whenever a person would go to town, they would purchase several nails so they would have enough to have a coffin built when they died. Nail making was also one of the very first cottage industries in colonial America.

Figure 8.4. Finished nailheader.

Before you can start making nails, you will have to make what is called a nailheader.

Nailheaders

Stock: 5/8 or 3/4 x 20 inch, square (mild steel)

Heat one end of the stock to a good yellow heat and, using the edge of the anvil, make a 90-degree bend 1 1/2 to 2 inches from the end. Reheat and upset the bent portion until almost flush with the stock. Now, take a punch slightly smaller than the stock to be used for the nails. For example, if you are going to use 1/4-inch stock for nail making, use a 3/16-inch punch for the hole, 5/16 inch for 3/8-inch stock and so on. I will usually, after the hole has been punched, give the top of the header a slightly convex face. I've found it makes it easier to flatten the nail head. (Figure 8.5.)

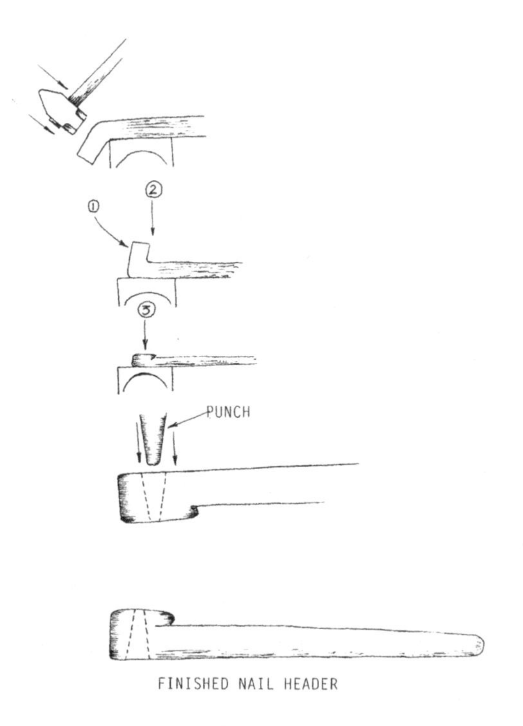

Figure 8.5. Steps in making the nailheader.

After the nailheader is finished, I drill or pound a small hole in the end of the handle to hang it on my anvil block.

Nails

Stock: 1/4 x 24 inch, round

Once you have your heading tool, you are ready to begin making nails. Take a 1/4-inch-round rod, heat the end to a cherry red, draw out to a point, and place the stock 1/4 to 1/2 inch above the drawn-out shoulder. On the hardy, hit it hard enough to cut most of the way through. Insert the nail in the heading tool and bend and roll it back and forth until it breaks off from the parent stock. Pound the head flat, cool in a bucket of water, and tap the nail out. (See Figure 8.6.)

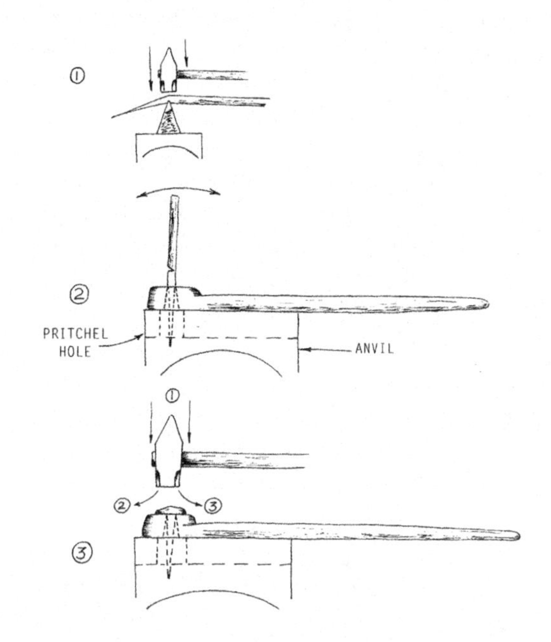

Figure 8.6. Making a nail.

After you do a few, you will develop a rhythm and procedure that is comfortable for you. With practice, a nail will take between 11 and 15 seconds. An advantage of this type of hand-forged nail is that if it needs to be clinched over, after it is driven through the wood pieces it is holding together, it can be done as shown in Figure 8.7. This leaves a very smooth and strong clinch.

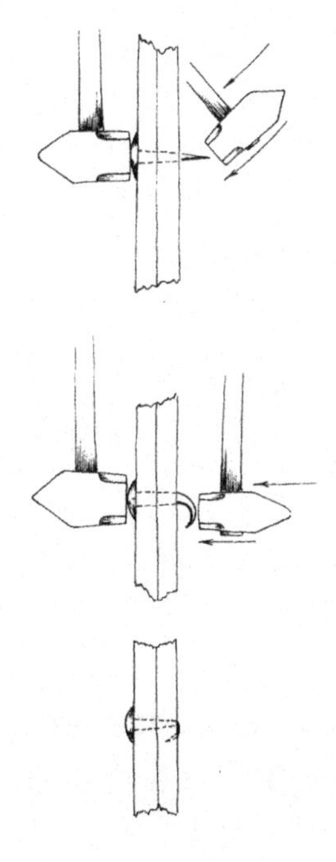

Figure 8.7. Clinching a nail.

For longer or larger spikes, heat the end of the rod and merely point it instead of drawing it out. Move up the spike 1/2 to 3/4 inch longer than the length you want and cut it off on the hardy. Now, heat the head end to a yellow heat and upset it by locking it in the vise and hitting it hard enough to upset it. Drop the spike through the heading tool and flatten the head just as with the drawn

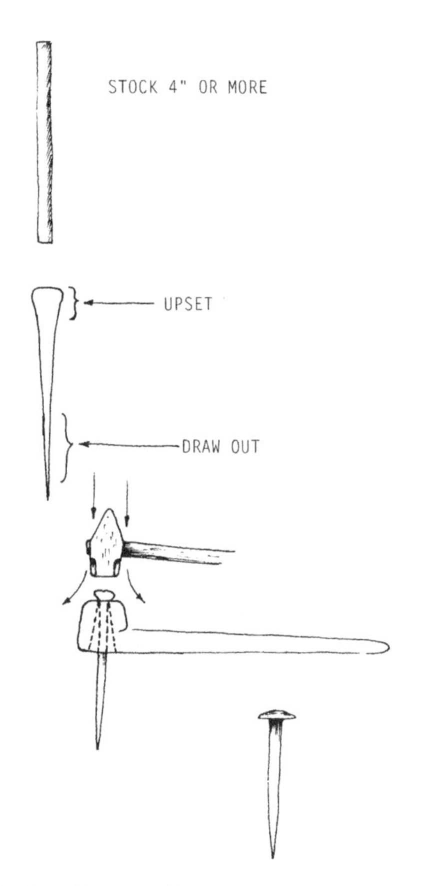

Figure 8.8. Forming a spike.

nail. (See Figure 8.8.) A variation of this process is used for making bolt heads.

HINGES

Stock: 1/8 x 1 1/2 x 18 to 24 inch (mild steel)

Hinges, regardless of whether they are made for trunks, cupboards, or corral gates, are all made in basically the same way. The hinges we will show here are simple strap hinges.

Figure 8.9. Completed hinge.

Heat the end of the stock and roll into a tight curl as shown in Figure 8.10. Be sure you start to roll the end very tight to get a smooth round curl. Do the same on the other end of the stock. Now, mark each piece into three equal sections with a white pencil or chalk. One end will be the female end, the other the male end. Make a cut with a hacksaw on each mark; then with a small chisel, cut the center piece out of one end (Figure 8.11) and the two outside pieces off the other end of your stock. You now have the two pieces that will be the hinge mechanism. Another method is to make the cuts *prior* to bending, as shown in Figures 8.12 and 8.13.

If you want a clean, tight fit for the pin, you can use a drill to finish the hole at this point. Now, determine the length of your hinge, mark and drill the holes, and cut the hinge from the parent stock. The pin for your hinge should be the length (width) of the hinge, plus the thickness of the pin. If the hinge is 1 1/2 inches wide, cut your pin at 1 3/4 or 1 7/8 inches long

Figure 8.10. Roll the hinge.

Figure 8.11. Knock out the center hinge portion for the female end.

Figure 8.12. Alternate method for making the male end.

for a 1/4-inch pin. Put the two pieces together, slide the pin in and peen one end over, turn the hinge, and peen the other end of the pin over.

Now that your hinges are done, take the nails you've made and use the hinges wherever they may be needed.

LATCHES

Latches can be made in a number of different ways, from the single hook and staple to the more elaborate thumb latches.

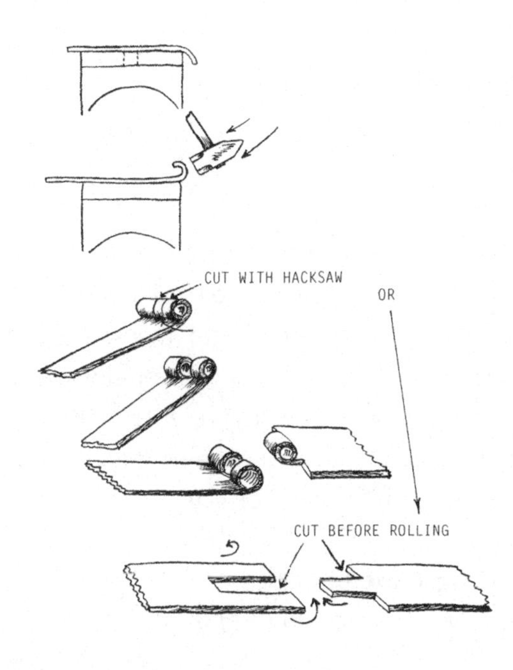

Figure 8.13. Two methods for a simple hinge assembly.

Figure 8.14. Spanish latch.

Gate Hook Latches

Stock: 1/4 or 5/16 x 2 to 3 inch, round or square (mild steel)

A basic gate hook latch is composed of two staples and a hook. The staples are small double-pointed, U-shaped nails pounded into the surface of the trunk, door, or post. It's simple yet effective. Make the eye as shown in Figure 8.15, measure the length of the hook, point the end, and bend. This may be used for doors or gates.

Sliding Bolt Latches

Stock: 1/4 x 1 x 24 inch (mild steel)

Another type of latching mechanism is the sliding bolt. (Figure 8.16.)

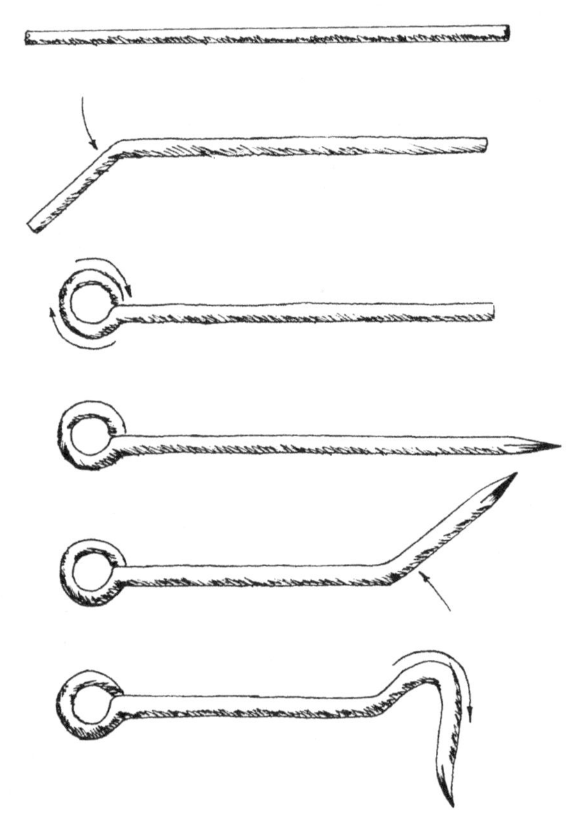

Figure 8.15. Forming a simple gate hook latch.

Taper one end of stock down as shown and make a simple curl. Measure in 3 to 5 inches and drill a 3/16-inch hole and countersink the back. Take a short piece of 1/4-inch round rod (3/8 inch long) and taper the tip a short way back. Place it in the 3/16-inch hole and peen over like a rivet. This is the "stop" to keep the bolt from sliding all the way out. Next you need to make three "keepers." To get a rough measurement for the keeper, take a piece of light wire, steel, or copper. Bend it around the bolt as a keeper would be used, mark it where it ends,

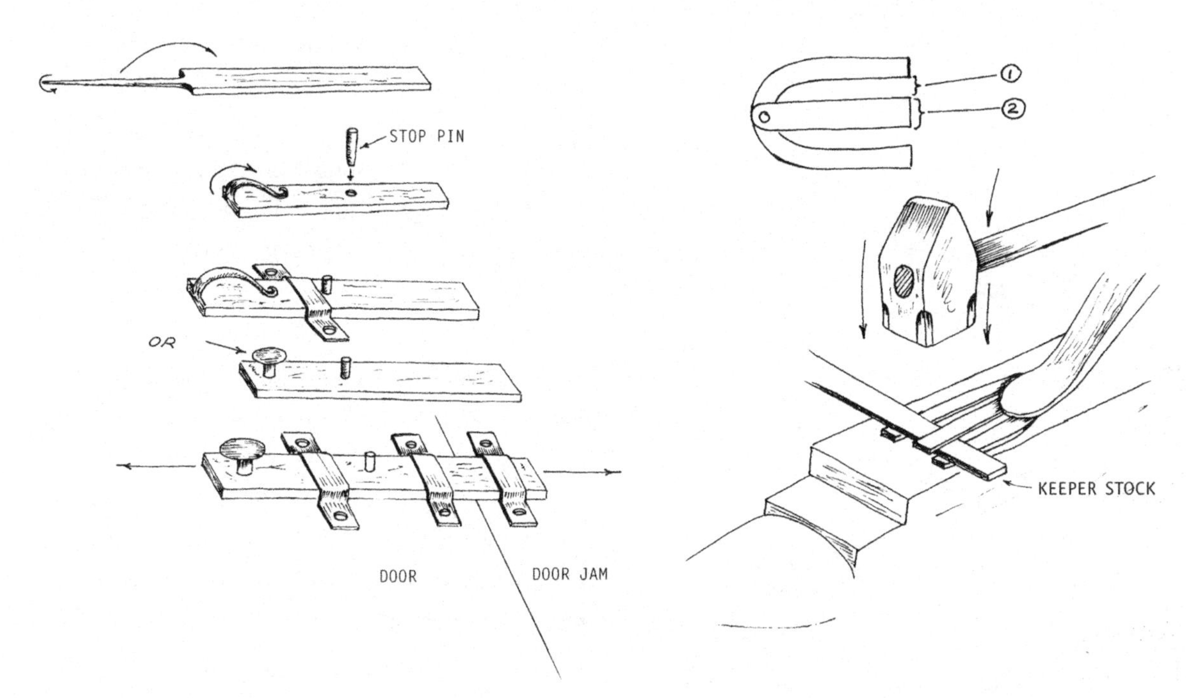

Figure 8.16. Sliding bolt latch assembly.

Figure 8.17. Jig for making keepers.

and then straighten the wire and measure. This will give a close measurement of the stock needed for the keeper. Add approximately 1/4 inch if using 1/8-inch-thick stock to compensate for the bends. A rule of thumb is to add half the thickness of the stock for each 90-degree bend.

A very simple jig can be made to facilitate the making of the keepers. What makes these a bit tricky is there are four right angle bends on each keeper. The jig is made as shown in Figure 8.17.

Spanish Latches

Stock: 5/8 or 3/4 x 1/4 x 12 inch

I refer to this as a Spanish latch because it is found throughout the Southwest region. However, I have also seen it used back east and in other parts of the country. This latch may be used for gates, doors, or trunks.

Heat one end of stock and, coming back approximately 1 inch, fuller down to a width of about 1/4 inch. Near the cut end, heat and punch a 3/8-inch hole. The 1-inch section on the end may be made into any number of designs, curls, leaves, or simple scrolls. I've even seen some with small animal heads on them. This will merely be the handle, so to speak, for moving the latch. From the fullered neck, come in 3/8 inch and punch (or drill) an oblong hole 1/4 x 3/4 inch

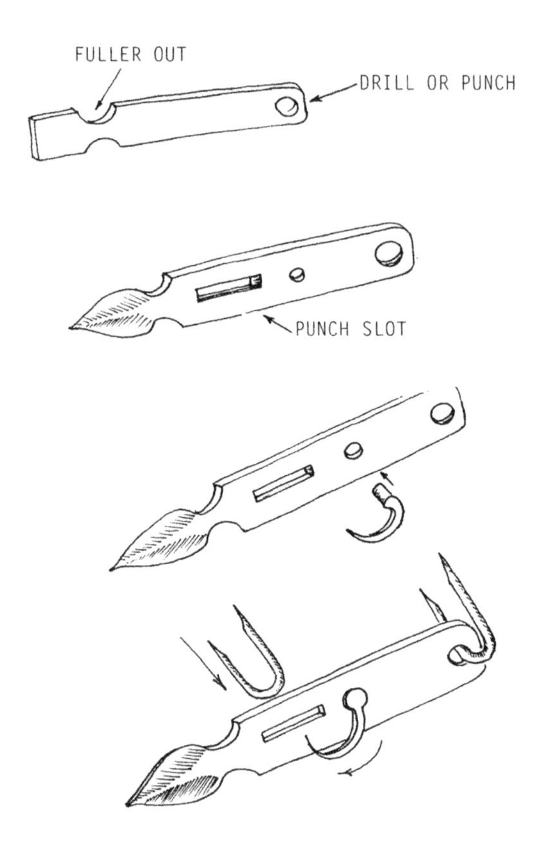

Figure 8.18. Spanish latch.

wide. The staple will fit into this hole for the hook. (See Figure 8.18.) Come back 1/2 inch from the end of the slot and drill a 1/4-inch hole in the center of the 1-inch stock. This is where the sickle-shaped pivoting hook will be mounted. To make the hook, take a piece of 1/4-inch round stock and draw out to a point about 2 inches long. Now, heat and make a 90-degree bend at the base of the point going 2 inches back. From the 90-degree bend, measure 5/16 inch and cut off. At this point you may leave the hooked portion round or you may flatten it. I have seen both styles. By flattening the hook slightly, especially at the base, the hook fits flatter against the body of the hasp. Before attaching the hook, measure back 2 inches and punch or drill a 3/8-inch hole for mounting the latch. Measure 3/16 inch from the back of the hole and cut off. If you want to do any decorative file work, now is the time to do it.

Take the hook, heat to a good red, place the base in the hole, and using a small pair of pliers, bend the hook so that it will engage the staple when it passes through the oblong hole. You may want to make the initial bend using the vise and small pliers and then test it and make any necessary adjustments. Once everything lines up to your satisfaction, attach the hook and peen over the base pin on the back. Remember, the hook assembly must move freely when finished.

This latch is more secure than the plain hook latch without the hassle of having to use a padlock. I have, however, seen this type of latch used in conjunction with a padlock by adding another oblong hole in the latch body. This type of arrangement is found on many old Spanish Colonial trunks.

Norfolk and Suffolk Latches

The basic difference between these two latches is the method of attachment of the thumb latch. The Norfolk has a back plate whereas the Suffolk does not. Both of these latches can be locked, so they work well for main door entrances.

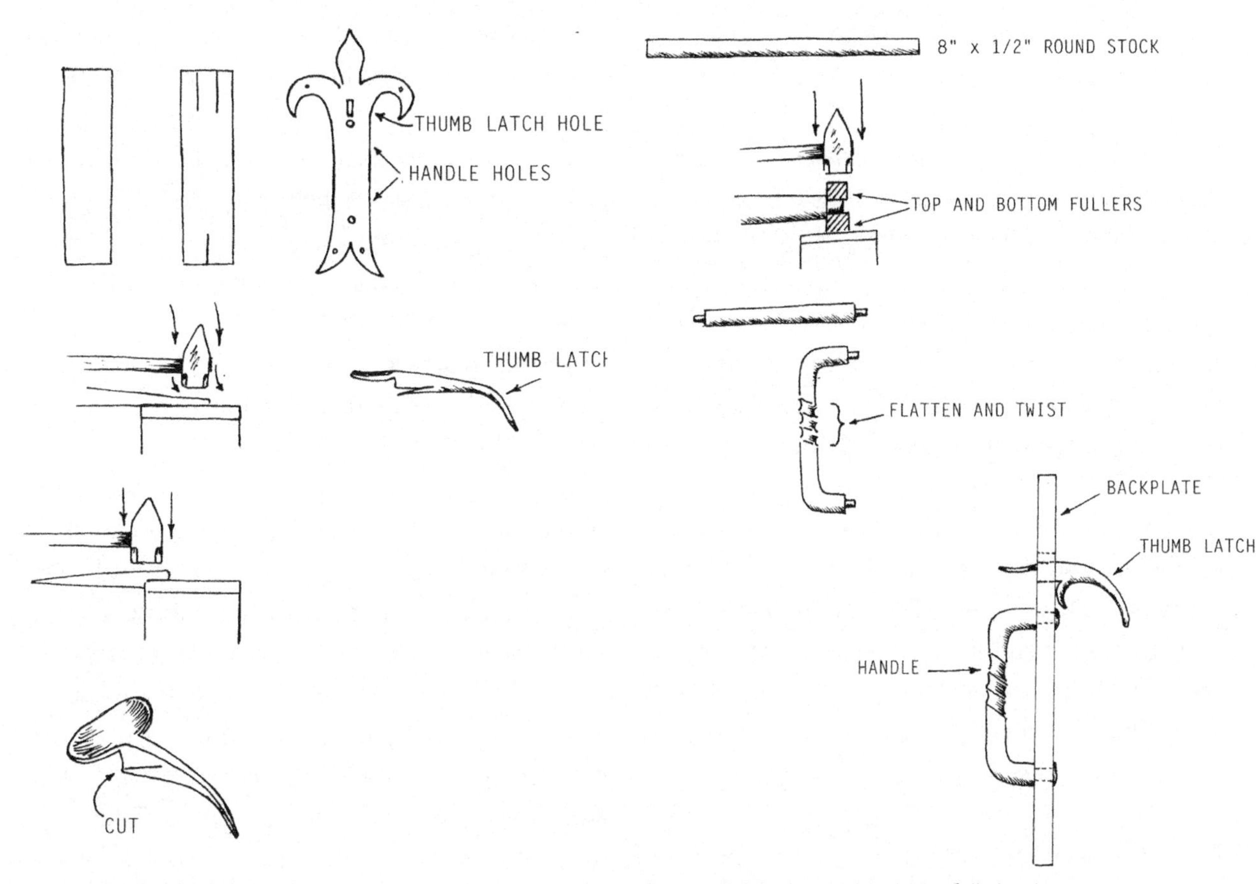

Figure 8.19. Norfolk latch.

Figure 8.20. Assembled Norfolk latch.

Norfolk Latches

Stock: Back plate: 1 1/2 x 7 inch x 12-gauge sheet metal
Handle: 1/2 x 6 inch, round
Thumb latch: 1/4 x 1/2 x 12 inch

Begin by making two cuts 1 1/2 to 2 inches long equidistant from each other on one end of the back plate stock. Draw out and curl the two outside pieces and shape the center piece. (Figure 8.19.) Bevel all edges. Next, with an oblong punch, punch the hole for the thumb latch.

Take the handle stock and on each end fuller a 3/8-inch-long tang for riveting to the latch plate. (See Figure 8.20.) Come back just past the tang on one end and flatten 3 to 4 inches. Bend to form a handle and notch the tangs to fit into the latch plate just below the oblong hole for the thumb latch.

For the thumb latch, heat 1 to 1 1/2 inches of the end of the stock and pound out flat for the thumb portion of the latch. Do this using the edge of the anvil so that you have a good sharp shoulder, as shown in Figure 8.19 on page 73. Draw the shoulder down a bit wider and taper the latch. At the base of the shoulder, saw a 3/4-inch cut with a hacksaw under the shoulder. This is the piece that will hold the thumb latch in place when completed.

Once all of this is done, heat the cut portion, slide through the slot, and bend down to hold the latch portion in place on the latch plate. This will be the back side of the latch plate and thumb lever. Take a piece of 1/4 x 1/2-inch stock and, using the end of the anvil as shown in Figure 8.19 on page 73, make a shoulder and draw out the end to form a finial. Measure back about 4 inches, cut and flatten, and then spread the end. This will be the pivot point on the door. It is *under* this bar that the thumb latch arm will make contact to lock and unlock the door for opening and closing.

Once the latch assembly is completed, you need to make the latch bar.

Latch Bars

Stock: 3/16 x 1/2 x 6 inch

Flatten approximately 1/2 inch of one end and drill a 3/16-inch hole in the center of the flattened area.

You now need a flattened staple and catch for the latch bar to engage. When making the staple, be sure to make it wide enough for the latch bar to have ample travel, or vertical movement, to engage and disengage the catch on the door jam. There are several ways of making the catch. The simplest is as follows: Measure in approximately 1 inch from the end and down 1/2 to 3/4 inch from the top and drill a 5/16-inch hole. Now, using either a hacksaw or small chisel, cut a slot in the catch plate for mounting to the door jam. The location of the holes will be determined by the thickness of the jam itself.

Suffolk Latches

Stock: 24 x 1/2 inch, square or round

Flatten and draw out one end. This will be the top of the latch. Just below this flattened area, take a rectangular punch or chisel and cut through. This will be where the thumb latch will be pinned. Neck down the section below this to approximately half of its original thickness for 4 inches, leaving a 1/2-inch piece at the end full thickness; cut from parent stock. (See Figure 8.21.) Flatten and spread the same as the top piece. Once both top and bottom finials are formed, drill a 1/4-inch hole in the center of each for attaching to the door. This will be the lower end of the handle to be attached to the door.

Make the thumb latch as for the Norfolk *without* the cut below the shoulder. Make sure this fits into the punched slot and moves freely. Once the pieces are fitted, drill a 1/8-inch hole from the side of the handle through the thumb latch. Insert a 1/8-inch pin and peen over, securing the thumb latch to the handle. The back side of the door latch assembly will be the same as for the Norfolk latch. You now have the completed Suffolk door latch.

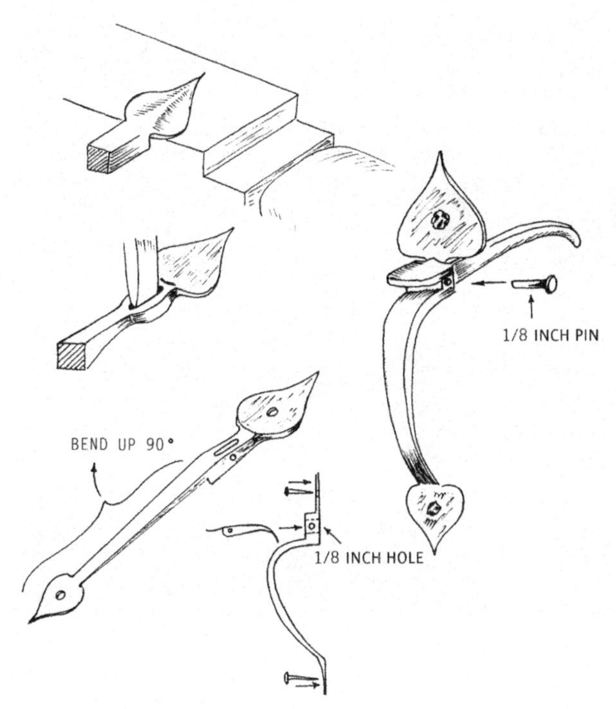

Figure 8.21. Suffolk latch assembly.

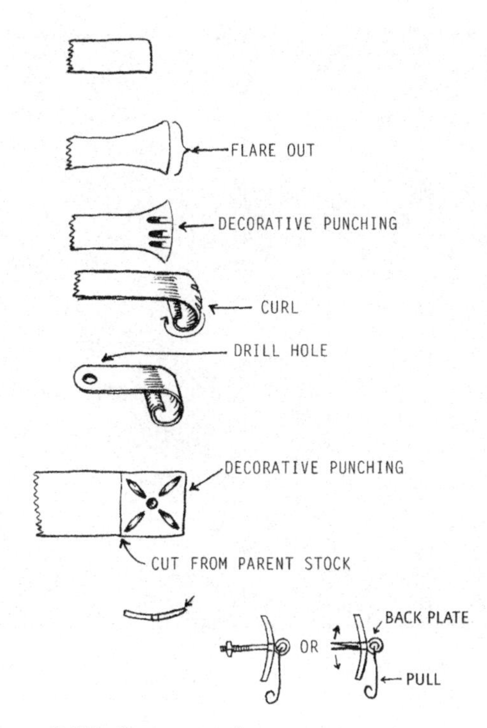

Figure 8.22. Drawer pulls.

DRAWER PULLS

Stock: 3/16 x 1/2 x 12 inch and 1/8 x 1 x 12 inch

An indispensable item for kitchen cabinet doors and drawers are drawer pulls. There are basically two types: drop pulls and handle pulls.

Drop Pulls

There are many variations to this design, so don't be afraid to experiment. These are made of three pieces: staple, backing or escutcheon plate, and the pull itself.

Heat the end of the 3/16 x 1/2-inch stock and flatten. You can either roll this into a simple curl or decorate it with punches and chisels. (See Figure 8.22.) Measure up about 1 1/4 inches and drill a 1/4-inch hole. Just past the hole, cut and round off the end with a file.

The next piece to be made is the back plate. Measure back from the end 1 inch and mark with a chisel. In the center of this 1-inch square, drill a 1/4-inch hole (if you are going to attach with a staple) or a hole the size of the shank (if you are using screw eyes).

If you're doing any embossing with punches or stamps, do as much as possible while the piece is still attached to the parent stock. Once you have all the work done, cut from the parent stock. Now,

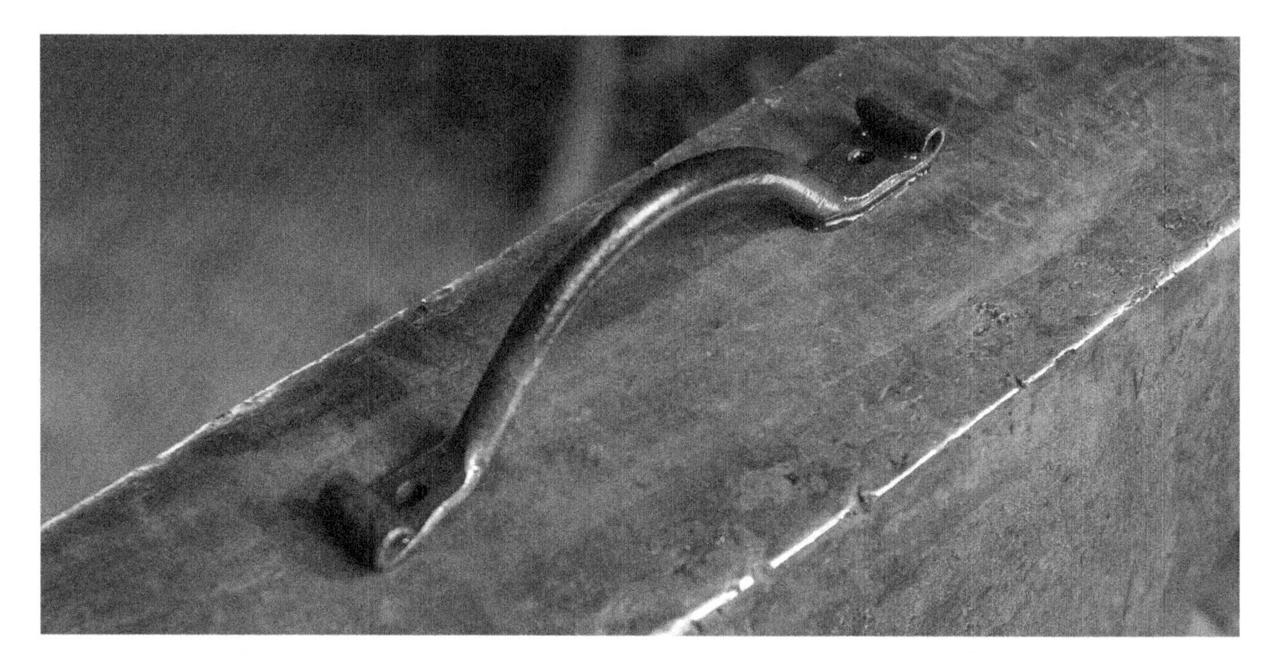

Figure 8.23. Completed drawer handle.

heat the piece, turn it upside down in a swage block, and hit the center to give the front a convex face. This helps "lock" the backing plate in place when the drop is tightened into place.

Drawer Handles

Stock: 1/4 x 12 inch, round

Like the drop pull, there are numerous variations of this design, but this is still basically a curved piece of steel large enough to get two or three fingers in, with each end flattened enough to allow for the drilling of a screw hole for attachment to the drawer or cupboard door.

Heat about 1 inch of one end, flatten using the end to form a shoulder, measure about 5 inches, and cut off. Using your tongs, heat and repeat the same process on the other end. Now, bend the handle enough to accommodate your fingers (Figures 8.24 and 8.25), heat and bend each end flat, true up the ends, and drill the holes. All that remains is to finish with a wire brush.

COAT HOOKS

Stock: 1/4 x 1/2 x 24 inch

A coat hook is a very utilitarian accoutrement to be sure, but it is one of the most useful pieces of hardware to be found around the homestead, whether it be the house, barn, or tack shed. The beam hook may be used as a coat hook, but I've found a wider hook is less likely to catch on clothing. Coat hooks also give you a chance to be creative with the finials.

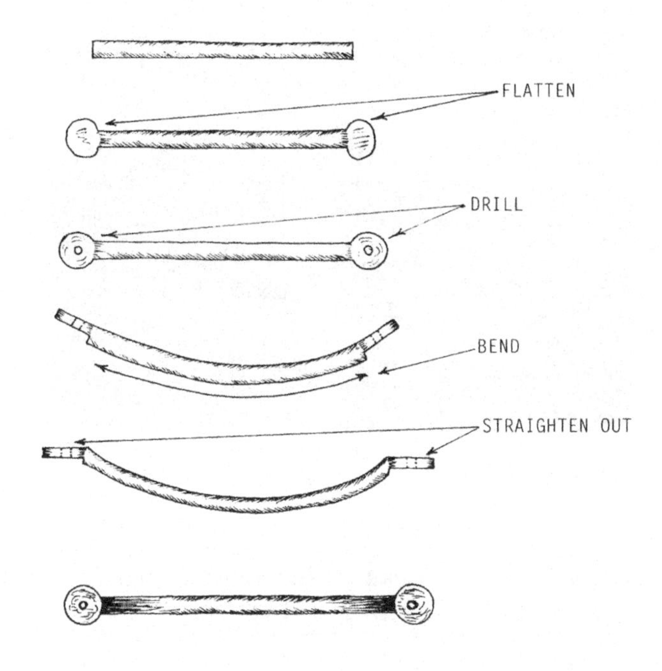

Figure 8.24. Drawer handle.

Figure 8.25. Make the bend of the drawer handle over the anvil horn.

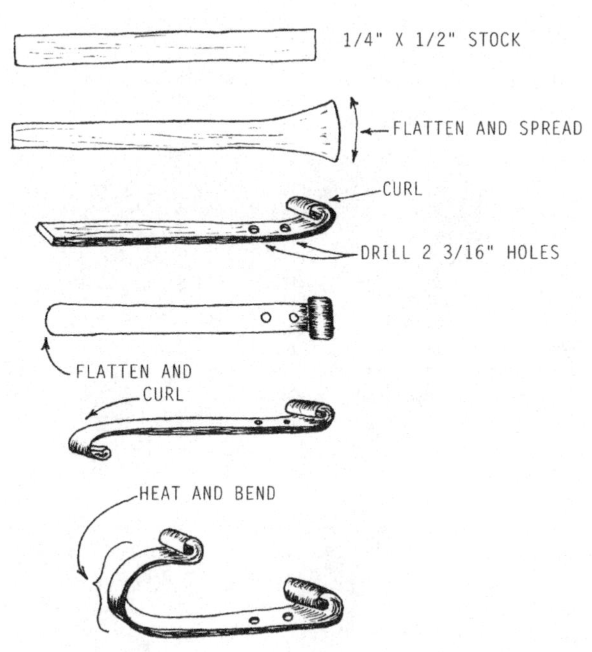

Figure 8.26. Coat hook.

This can be anything from a simple curl to a split curl, leaf, or the more elaborate animal heads. (See Figure 8.27 on page 78.)

I normally begin with the finial since I've found it easier to hold onto the parent stock when doing leaves, split curls, and especially the animal heads. Once the finial is finished (I usually drill two holes approximately 1 inch apart just below the finial), measure down about 6 or 7 inches from the finial and cut. At this end, flatten and then curl. Remember to curl *away* from the front of the finial. (Figure 8.26.) Now, heat the lower three inches, cool just the *tip*, and bend the hook portion either over the anvil horn or on a jig. It is important here not to bend the hook too far back on itself if you

Figure 8.27. A horse head finial on a coat hook.

are using this as a clothing hook since it will tend to catch on clothing.

TOWEL BARS

Stock: 1/2 x 26 inch, round

This type of towel bar may be used for regular towels or paper towels since it is open on one end. Other than the type of finial you want on the wall mount, the rest is just a series of bends with a curl on the end to keep the paper towel roll from coming off. It also gives the piece a more finished

Figure 8.28. A ram head finial on a towel rack.

appearance. The toilet paper holder is made the same way only on a smaller scale.

From one end, come in 5 inches and mark. This is the portion that attaches to the wall and will have the finial. Measure down another 5 3/4 inches and mark; your next mark will be 2 inches from the last mark. For the last mark, measure another 11 1/2 inches. From that mark, measure 1/2 inch and cut off any remaining stock. This 1/2 inch will be the curl on the end of the towel bar.

Now that all measurements are marked with chalk or pencil, go back and mark again using the hardy. It is important to notice the direction of each bend and to make the mark on the *inside* of the bend.

Begin by heating the end and making the finial. One of the simplest ways is to merely flatten and draw out the end and roll it into a graceful curl. I get more requests for this type of finial than any other. Once

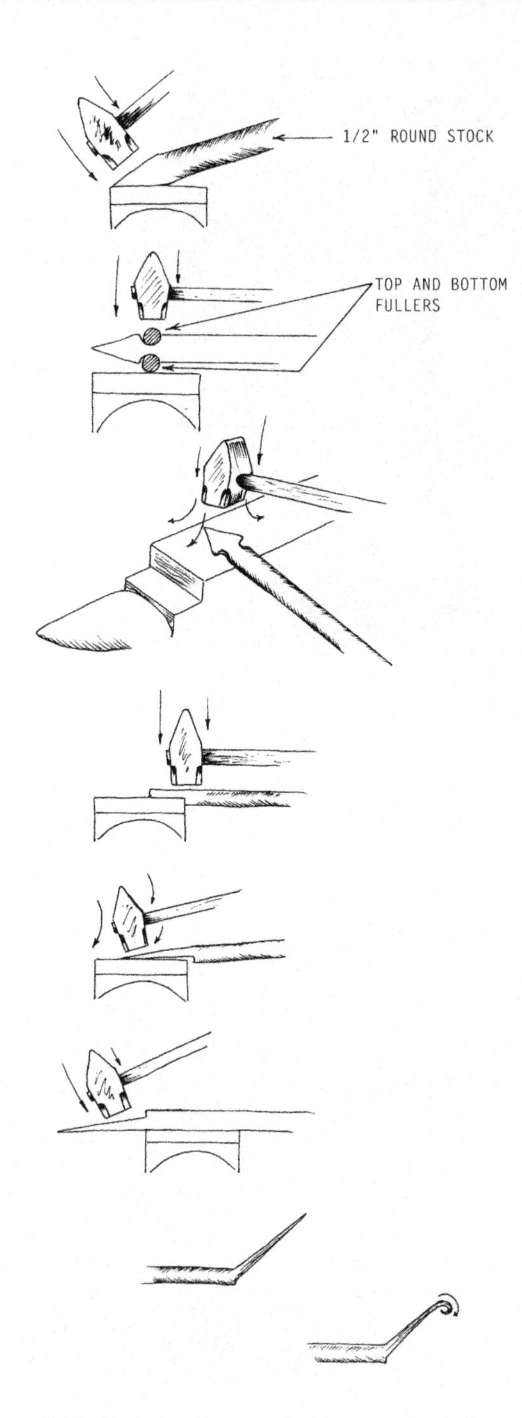

Figure 8.29. Forming the end finial on the towel bar.

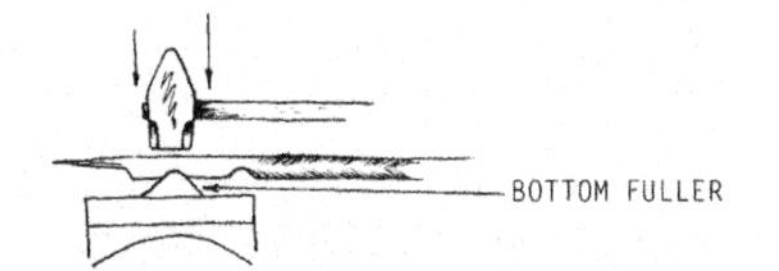

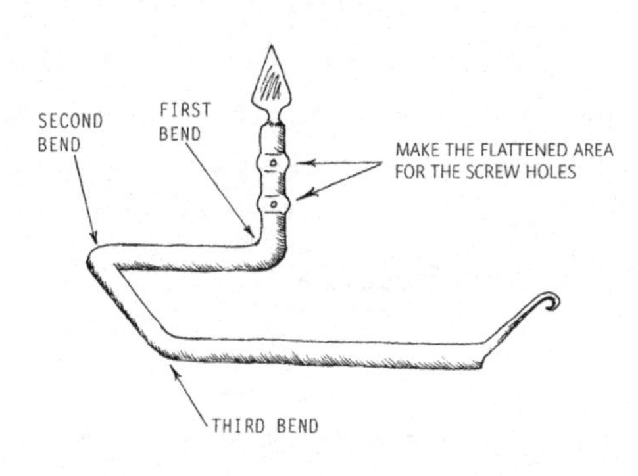

the finial is done, heat and flatten an area below the finial for approximately 2 inches. This is where you will drill the holes for mounting. Make them about 1 inch apart and 3/16 inch in diameter. Now, heat the other end and draw out the 1/2-inch length marked for the curl. Draw out to a good tapered point, bend up slightly, and curl the end. (See Figure 8.29.)

All that is left now is to heat at the marks and make three 90-degree bends. Your towel bar is ready to mount on the wall.

As was mentioned earlier, the toilet paper holder is made in the exact same way using smaller stock (1/4 x 1/2 inch flat or 3/8 inch round) and smaller dimensions. The most important dimension here is the length (width of the toilet paper roll).

CANDLEHOLDERS

Lighting fixtures are a necessity in any home and take many and varied forms. For our purpose here, we will cover the simple table candleholder and wall mount. Again, you are only limited by your imagination as to the design. There are two basic types of candleholders: the socket and the spike.

Figure 8.30. A ram head candleholder.

Socket Candleholders

Stock: 3/8 inch x 2 feet, round

Heat and flatten one end back about 1 inch. Cool, and then drill a 3/16-inch hole about two-thirds of the way back on the flattened area. You may want to roll the tip of the flattened portion slightly as a decorative touch. (See Figure 8.31.) Now, begin heating and bending it into a circle with a flattened area in the center. Think of it as a snake coiling up, and the flattened portion as the snake's head. Once you have made a complete circle, measure about 4 to 5 inches and cut. This portion will be the handle. Bend this area up and out and form into a circle or finger ring.

Take a 2- to 2 1/2-inch square piece of 16-gauge sheet metal and cut or file into a circle, dish slightly, and drill a 3/16-inch hole in the center. This will be the drip pan. It is now time for forming the socket that will actually hold the candle.

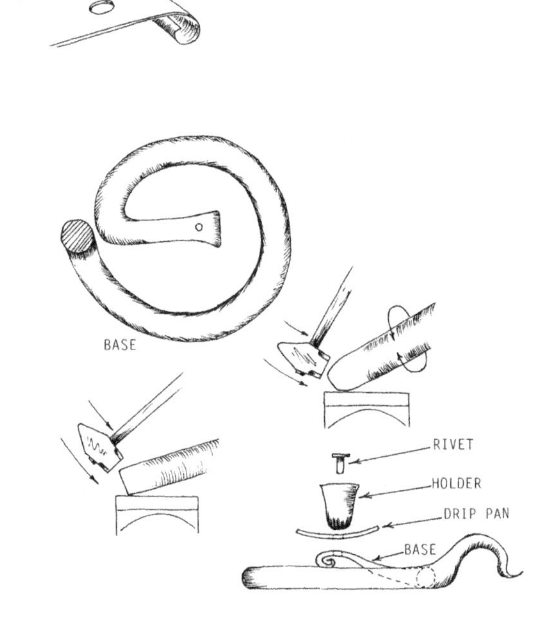

Figure 8.31. Forming the socket holder.

Take a 24-inch piece of fairly heavy walled pipe, heat, and flare one end as follows: heat the end,

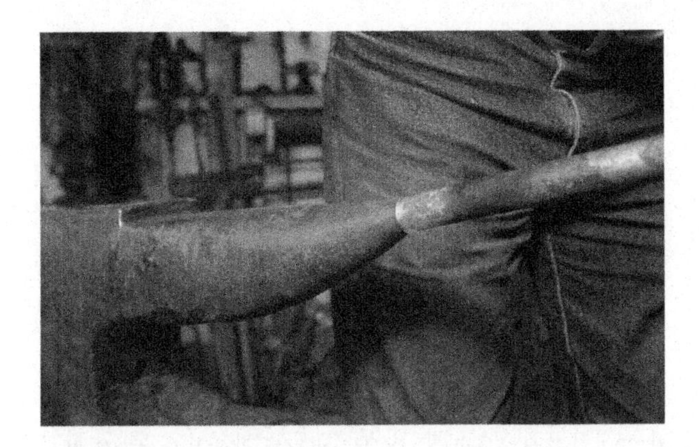

Figure 8.32. Flare the top of the socket.

Figure 8.33. Close the base of the socket.

Figure 8.34. Socket and drip pan are ready to be riveted to the base.

slip it over the end of the anvil horn or bick, and while turning the pipe, give it several sharp blows on the other end. (See Figure 8.32.) Turning the pipe is done to keep the flare even. You'll notice the thin wall of the pipe cools fairly rapidly, so this process will take several heats depending on how much you want it flared.

Measure back $1\frac{1}{2}$ to $1\frac{3}{4}$ inches and cut. Heat the end you have cut to a good red or red-yellow heat and begin with a light hammer, constantly rolling the piece back and forth while hitting it to bring the open end together. (See Figure 8.33.) This will take several heats as well. Remember to keep rolling it back and forth so you don't get any flat spots. Take your time.

Once you have rolled the end closed, drill a $\frac{3}{16}$-inch hole in the bottom. Rivet the drip pan and socket to the flattened area of the holder. You may need to do some adjusting to make sure the whole thing sits level and straight. (Figure 8.34.)

Spike Candleholders

Stock: $\frac{3}{4}$ x 1 x $\frac{1}{4}$ x 24 inch

This will be a wall-mounted candleholder. Heat and flatten one end. Drill a $\frac{3}{16}$-inch hole as with the socket holder. Measure back 3 to $3\frac{1}{2}$ inches, heat, and bend in a hooklike shape with about a 2- to $2\frac{1}{2}$-inch radius. For a simple finial at the top, measure up how far you want, cut, and flatten. Before rolling or curling, take the cross peen of your hammer and texture the area. It adds a decorative touch to the finial. Roll the finial in a large curl, or you may want to leave it straight with the textured and flared area flat against the wall. (See Figure 8.35 on page 82.)

For another type of finial, use the split and curl technique. After you have cut the piece, mark the center of the piece for about 2 to 2 1/2 inches. This is best done using a center punch. Mark in three or four places. Heat and split. Once you have it split, spread the split and begin drawing both sides out over the horn into a half circle. Drill two holes in the wall mount.

The spike type of candleholder works best with larger candles: 1 1/2 to 2 inches in diameter or larger. Cut out a drip pan as with the socket candleholder and drill a 3/16-inch hole in its center. Take a piece of 1/4-inch round stock, long enough to hold onto while you work it, heat the end, and draw down to a sharp point (1 inch or a bit more). Come back to the base of the point, where you are at the full 1/4-inch thickness, and add approximately 3/8 inch or enough to go all the way through the drip pan and the body of the candleholder to be riveted over.

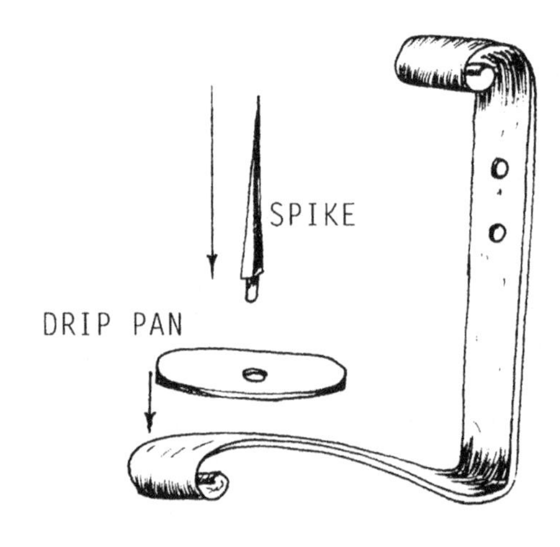

Figure 8.35. Assembly of spike candleholder.

Just below the base, file the stock so it will just fit through the 3/16-inch hole. Now, take the spike and lock it in the vise, point down, just below the shoulder, slip the drip pan on over the bottom of the spike, and then do the same with the candleholder body and rivet all three pieces together. Use a light ball-peen hammer so as not to deform the spike. Once everything is riveted together, remove from the vise and make sure everything is aligned correctly.

This same basic procedure may be used to make multiples of the wall mount to hold two or more candles.

FIREPLACE EQUIPMENT

There are three basic tools for the fireplace: the poker, the shovel, and the broom. (See Figure 8.36.) Another piece used by many is a set of tongs for positioning the logs in the fire. You will also need a wall mount or stand to hang everything on.

As with the coat hooks, there are any number of handles and finials that may be used, limited only by your imagination. The stock you use will depend on the size of your fireplace. If you have a small kiva type, 3/8-inch round or square stock will work just fine; for a larger, deeper fireplace, 1/2 inch works better. For the purpose of this book, I will be using $^3/_8$-inch stock.

Pokers

Stock: $^3/_8$ x 32 inch, square

You will be making the handle first. The handle described here is easy to make and very functional. Start by flattening one end for about an inch or so and then curling it to an open curl. Make sure your forefinger will fit comfortably inside the curl.

Now, measure back about 5 inches and heat that area. When it reaches a good red, bend it in the *opposite* direction as the finger curl. This may be done over the horn or using a jig. Be sure to keep a radius at the bend. I usually use a $^3/_4$- to 1-inch-radius bending jig. (See Figure 8.37 on page 84.)

With the handle done, you will now move onto the working end of the poker. There are two basic styles used. The most common is the two-prong type: one prong straight, the other hooked. The other type is just a plain hook, like your forge poker.

The most practical two prong is the easiest to make. I say that because some I have seen are so outlandish and unwieldy as to be totally impractical.

Begin by flattening approximately 2 inches of the end of the poker. Now, with your chisel, split the flattened area about $1^1/_2$ inches and spread the two pieces. Draw one out using the horn of the

Figure 8.36. Fireplace tools and stand.

anvil, which will give it the hook shape. (See Figure 8.37 on page 84.) Draw the other piece out on the anvil face. This will be the straight part of the poker. After completing this portion of the poker, you may want to put a decorative twist in the middle of the poker. Merely heat the desired area, lock one end in the vise and, using an adjustable wrench, make your twist. Be sure the poker end is properly aligned with the handle and that the

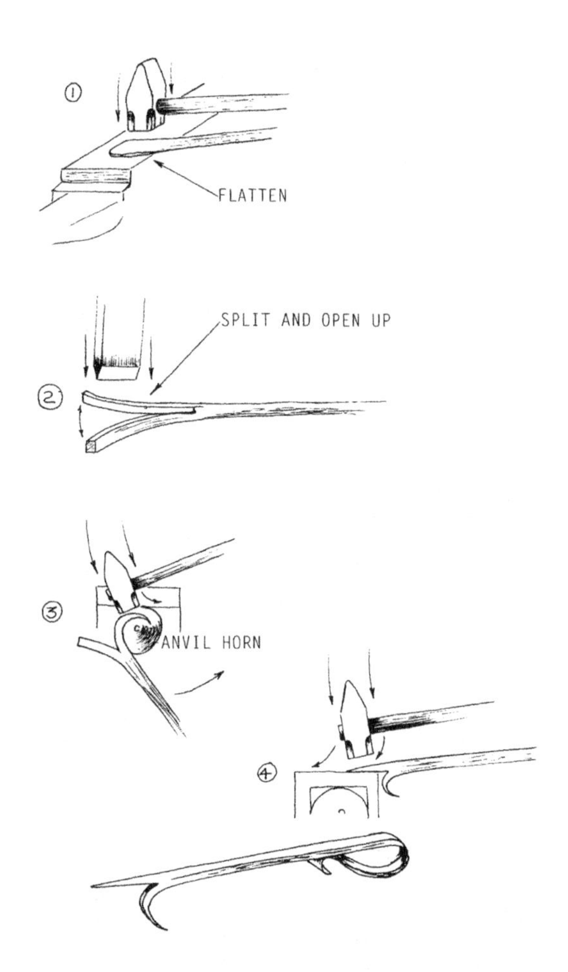

Figure 8.37. Fireplace two-prong poker.

shaft is straight. This may be accomplished by locking the twisted area lengthwise in the vise and adjusting either end of the shaft by tapping lightly with a hammer. If you put a twist in one utensil, it should be followed through on all pieces, including the stand.

As mentioned previously, the other type of poker is the regular hooked poker that is used with the forge tools. The two-prong type is the more practical of the two for moving logs around in a fireplace.

Shovels

Stock: 3/8 x 24 inch, square, and one piece 16-gauge sheet metal, 8 x 8 inch

Before making the handle, you must make the shovel itself. Take the piece of sheet metal, measure in 1 inch from the side on both sides, and draw a line from this point to the corner, as shown in Figure 8.38. Cut along the lines with a chisel or tin shears. This will give you a sort of trapezoid shape. On each side, measure in 3/4 inch and draw a line from top to bottom. On the narrow end, measure in 3/4 inch and draw another line across the top. Take a chisel, or tin snips if you prefer, and cut in 3/4 inch on the side lines at the top. Now using either the heel of the anvil or vise, bend up the back of the shovel 90 degrees. Bend up each side the same way. You will notice two tablike protrusions sticking out the back of the shovel. Bend these around the back of the shovel.

Make the handle as you did for the poker. Now, lay the shovel and poker side by side. Place the sheet-metal shovel so it is even with the tip of the poker, and mark on the shaft of the shovel where the back of the shovel is. From that mark, measure down 1 1/2 inches and cut off any excess. Using the end of the anvil to form a shoulder on the shaft, draw out to 1/4-inch thickness. Drill two holes on this drawn-out portion, countersinking the back. Line the shovel up on these holes, mark, and drill two holes. Remember, the rivets will go through

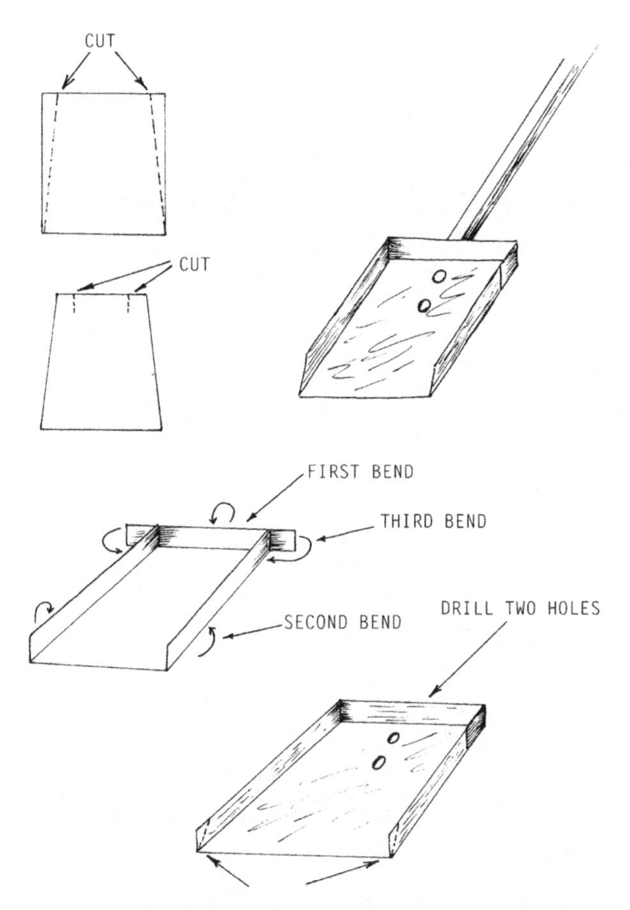

Figure 8.38. Shovel.

on the shovel side and be peened over on the back or shaft side.

Brooms

Stock:

Handle: 3/8 x 24 inch, square

Broom: 1/4 x 1x 12 inch

Ideally, the best way to make a fireplace broom is to locate a broom maker who will make the

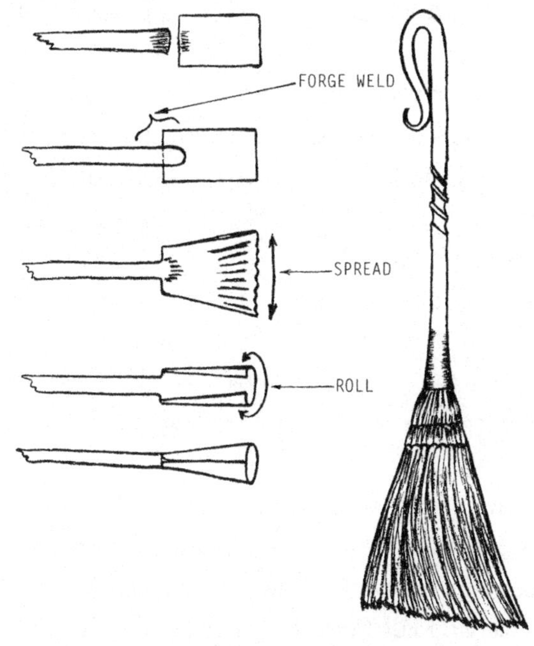

Figure 8.39. Making the socket for the broom.

broom bristles right on the shaft of the broom. (At the end of the book I've listed one that I know.) A lot of people prefer to use a large whisk broom.

The broom handle should match the handles of the poker and shovel. Measure down about 8 inches on the shaft and cut. Take a piece of 1/4 x 1 x 12-inch stock and flare one end to a triangular shape as with the socket chisel. (See page 102.) From the base of the triangular piece, measure back about 1 inch and cut. Heat this area and fuller out this end in the middle in preparation for forge welding to the handle. Now fuller the tip of the handle out, heat both pieces to a dull red, flux, and return to the fire, bringing both pieces to welding heat. Finish the weld on the anvil. I usually use a small "closet broom" and cut the handle down to where it will fit in the socket. At this point, I drill a small hole in the socket so I can screw a small screw into the handle of the broom. By using this method, the broom head itself can be easily replaced.

Hoes

Stock: 1/4 x 1 x 36 inch

To make a hoe, we combine several basic steps: drawing out, bending, and riveting.

From an end, measure back about 3 inches and, using a fuller or edge of the anvil, draw down to about 3/8-inch wide. Now, measure back 3 to 3 1/2 inches and cut from the parent stock. This will be spread and drawn out to form the socket. Once you have spread, drawn, and rolled the socket, it is time to prepare the end that will be riveted to the hoe blade. (See Figure 8.40.) On the shank, measure back approximately 1 1/2 inches and mark. Place the mark on the sharp edge of the anvil and pound the piece enough to flatten it slightly. At the same time, form a shoulder with the edge of the anvil. Drill two holes in the flattened area for the rivets. Once the holes are drilled, turn the piece over and countersink the holes on the opposite side of the shoulder. This will be the side that is riveted over.

Measuring back from the shoulder 1 inch, heat, and make a right-angle bend toward the shoulder. You are now ready to make the blade. Take a piece of 1/8-inch stock and cut to the desired shape. Position the blade on the shank, making sure the top is tight against the shoulder. Mark the holes and rivet. You will probably want to pin the handle/socket, but this can be done once the handle is installed.

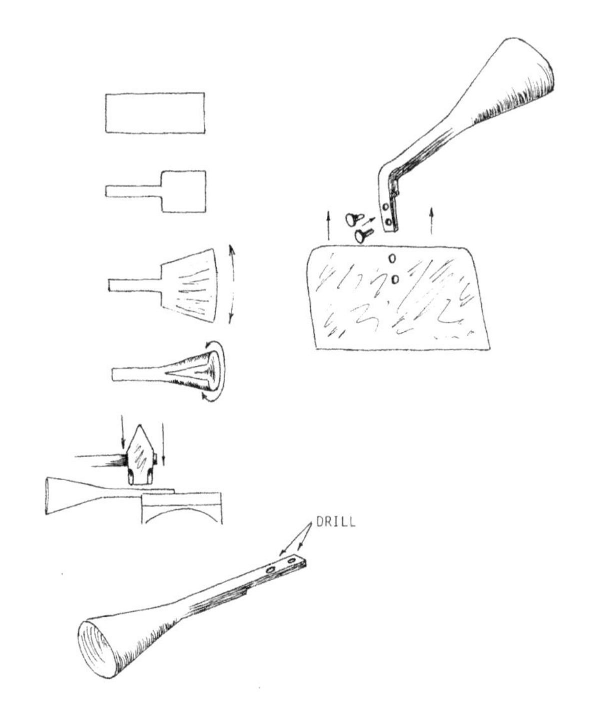

Figure 8.40. Steps for making a hoe.

Variations of this same technique can be used to make a small garden rake, trowel, or weeder.

CROWBARS

Stock: 5/8 x 30 inch, square

Crowbars and spud or pry bars find numerous applications around the homestead and ranch and can be made in various sizes. From a practical standpoint, two feet is the minimum length for either type.

When making the crowbar, remember you are giving the stock a quarter or 90-degree turn before flattening either end to give greater strength for prying. In the cross section of Figure 8.41, you can see why.

For the nail-pulling hook end, begin by first flattening the end with a fairly thick or blunt one-sided taper. Again, this is for strength. Once this is done, reheat and split the end back about 1 to 1 1/2 inches. Spread slightly as with the hammer. Next, heat the other end and, giving the stock a 180-degree turn, do another rather blunt one-side taper as shown. This will be the prying end. Now, reheat the same end, and laying the base of your taper over the end of the anvil, give it several light blows to put about a 30-degree angle on the end. Sharpen at a fairly sharp angle.

To make the hook, heat the nail-pulling end about 4 to 6 inches, depending on how big of a hook you want, and bend it over the horn or a bending jig placed in the hardie hole. Make sure both ends are aligned with each other and cool off.

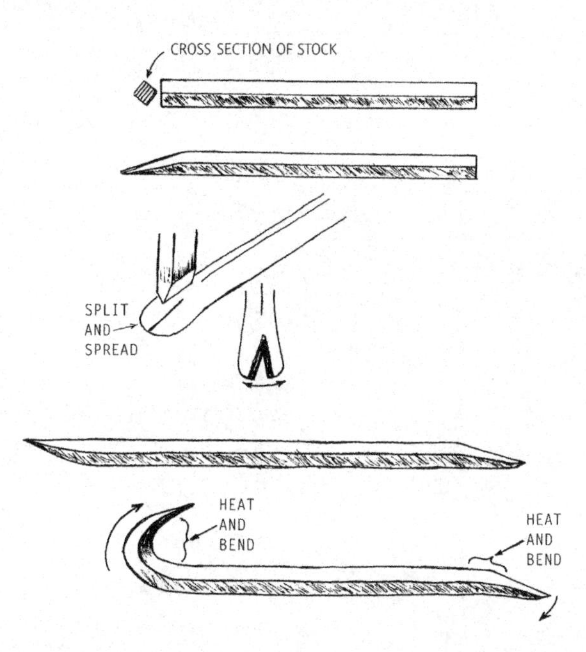

Figure 8.41. Crowbar.

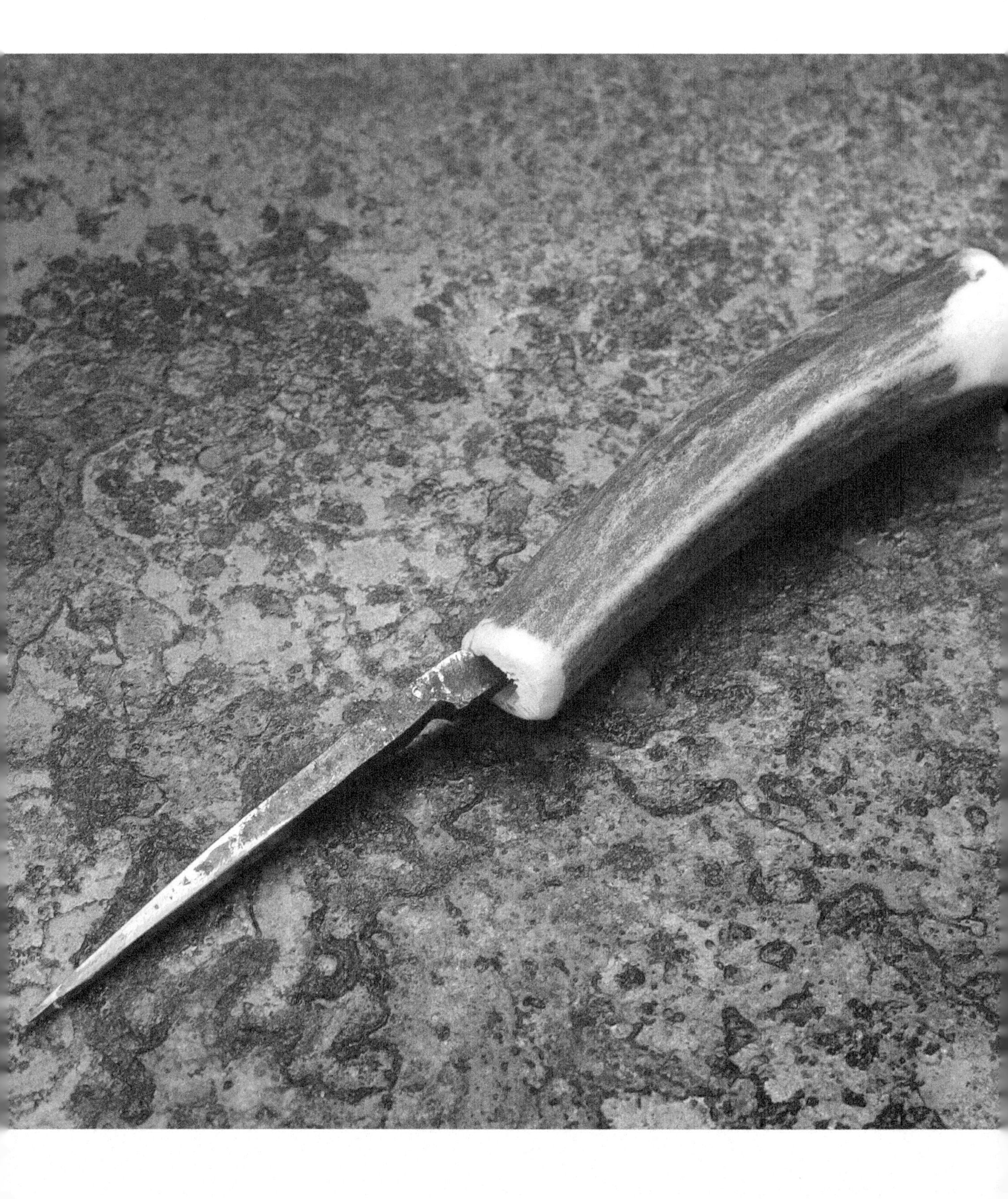

9 Toolmaking: Basics

AWLS

Stock: 1/4 or 3/16 x 10 inch, round or square

The awl is a useful tool around the homestead for everything from fixing leather harnesses or clothing to starting small screws in wood. The type I make is referred to as dogleg style.

Draw one end of your rod out to a long, narrow point. Now, measure back about 3 inches or so and cut off on the hardy. Draw out the other end to a good point. Next, in about the middle, hit the opposing sides 3/8 to 1/4 inch apart over the end of the anvil. (See Figure 9.2.) This dogleg will keep the awl from sliding through the handle and into your hand. Handle material may be a piece of wood or deer antler. Merely drill an undersized hole an inch or so into the handle material, and while one end of the awl is held in the vise, drive the handle down on the awl by lightly tapping it with a hammer.

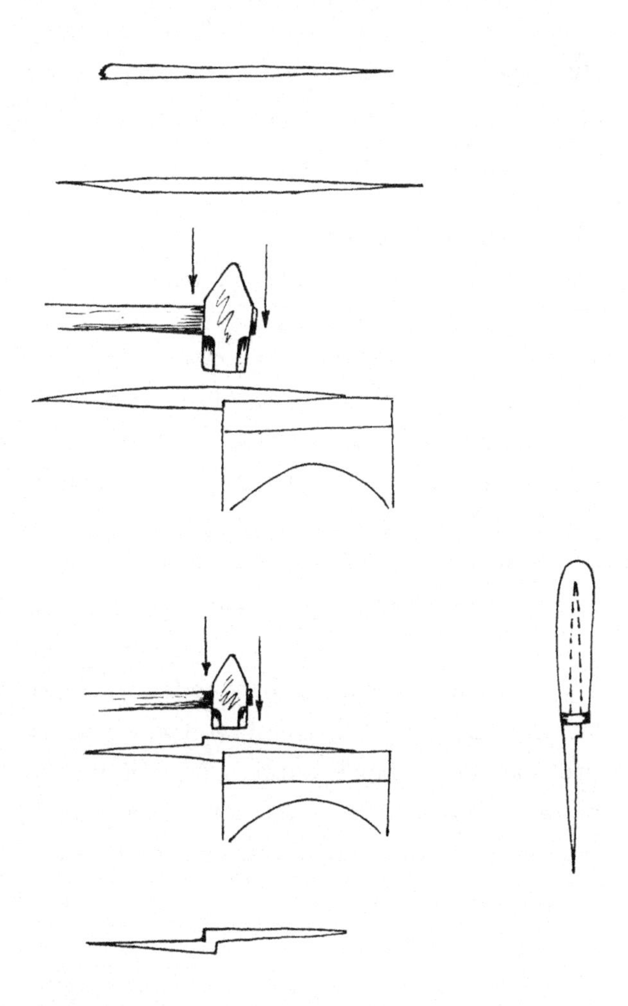

Figure 9.1 (facing). Dogleg awl with antler handle.
Figure 9.2 (above). Making a dogleg awl.

SCREWDRIVERS

One of the simplest yet most useful tools to make is the screwdriver. There are two types we will discuss here. One is the small, lightweight household screwdriver; the other is the heavier shop screwdriver.

Household Screwdrivers

Stock: 1/4 x 24 inch, round or square

Make the handle first; this description will be of a teardrop design. Draw out to a point, curl, and make the first bend about 4 inches up from the end of the curl. Make the second bend in on itself to form the teardrop shape; then flatten the other end. (Figure 9.3.) File to shape the end. If using high-carbon steel, bring to a red heat, quench, and draw to a dark straw color.

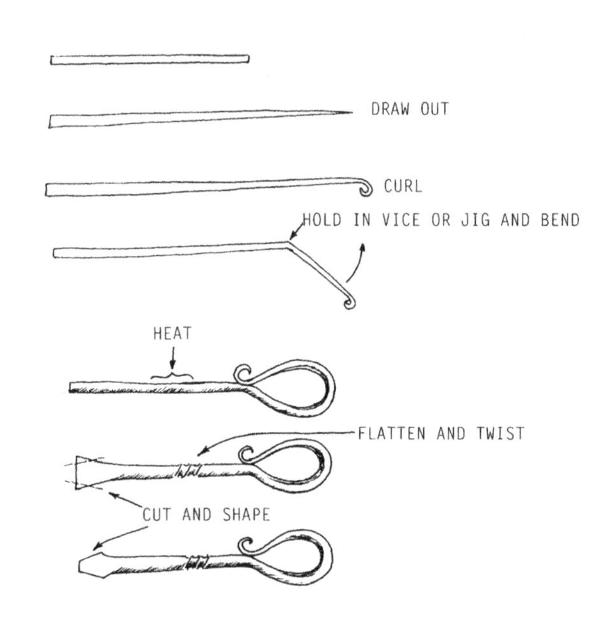

Figure 9.3. Household screwdriver.

Shop Screwdrivers

Stock: 3/8 (5/16) x 24 inch, round or square

To begin, flatten the handle, as shown in Figure 9.4, and then draw out the middle of the screwdriver shaft, leaving the end at the thickness of the original stock for about 1/2 inch. Flatten the end and finish as you did the small screwdriver. Now, drill three 3/16-inch holes in the handle. Glue and pin two pieces of wood and finish out to your personal preferences.

You now have some of the basic tools and accoutrements for your home and shop. You will notice that the more complex pieces do not require more complex procedures but rather merely combine a number of basic procedures to achieve the end result.

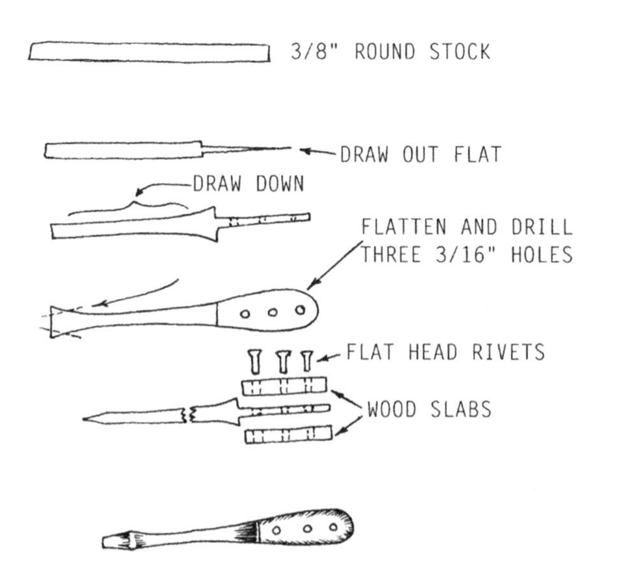

Figure 9.4 (above). Shop screwdriver.
Figure 9.5 (facing). Household, left, and shop screwdrivers.

10 Toolmaking: Forge Welding and Tempering

In this chapter, we will be covering more complex tools and will be using more advanced procedures, such as forge welding and tempering, in conjunction with the basic procedures with which you are already familiar.

Here are some things to keep in mind:

1) Before beginning the procedure, understand it and take your time.
2) Plan ahead. Know the tools you are going to need and have them handy.
3) Don't panic. Forge welding and tempering can be challenging when you first start, but don't get discouraged. You can do it.
4) Practice, practice, practice, practice . . . well, you get the idea.

FORGE WELDING

Forge welding is not magic, as some would have us believe, nor is it a special innate skill possessed by a chosen few. It is merely a method for joining two pieces of iron or steel by using heat from a coal or charcoal forge. If there are any "tricks," it's knowing your steels and knowing when you have reached the welding heat. Both of these are acquired with practice, so they are hardly a trick. I say all of this to allay any fears or mystical misconceptions people have about forge welding. It is a mechanical operation perfected through trial and error and practice. It's no more difficult than bricklaying, cabinetwork, or pottery.

Figure 10.1 (facing). Forge welded chain link.

Forge welding, in its simplest form, proceeds as follows:

1) Have a clean fire, which is one without clinkers. This is the most important step.
2) Heat the metal.
3) Flux the metal.
4) Reheat to welding heat.
5) Remove the metal from the fire and pound together.

CHAIN LINKS

Stock: 1/2 or 3/8 x 1/4 x 11 inch, square

A flat chain link works well as a beginning forge welding project. The reason for this is that when it is time for the actual weld, all pieces are in place and you don't have to fumble around trying to place two separate pieces together before grabbing your hammer and hitting it. An additional advantage to doing a chain link or oblong ring is that it is much easier to localize the heat at the point of the weld.

Heat the rod in the middle and bend in a U shape. (See Figure 10.2.) Flatten one end of the U on one side, and then turn it over and flatten the other end on one side as shown.

Now, heat both tips for an inch or so and, using the tip of the horn of the anvil, bend both ends toward each other as shown in the illustration. If you have lost your red heat at this point, reheat the ends, flux them, and place them back in the fire.

The fire is the single most important part of forge welding. The fire does the work; you merely manipulate the tools. All clinkers should be removed prior to the actual welding heat. These will show up as dull gray or black in the fire itself, resulting in dead spots. They will feel like rocks or very hard pieces of coal. For those just starting out with forge welding, it is best to let your fire sit idle for a few minutes prior to your welding project. This gives the clinkers a chance to cool down somewhat and solidify. When you start your blower, the clinkers will show up very clearly and can be lifted out easily with your straight poker or shovel. If your fire is burning cleanly, it will be at a white heat in the center where the metal is to be heated.

Now that the piece to be welded is shaped and fluxed and the fire is clean, the actual welding process can begin. Place the ends of the link deep

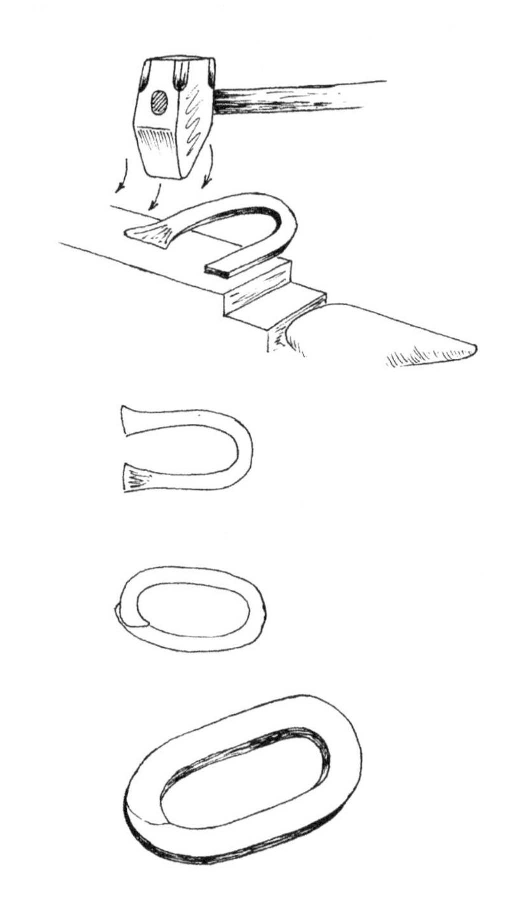

Figure 10.2. Forge welding a chain link.

in the center of the fire and begin with a relatively slow air blast if you have a hand-pumped bellows or a variable-speed blower. You will have to turn the piece several times while in the fire because you must have an even heat.

Both pieces must come to welding heat at the same time. As the metal heats to the red and orange stage, the flux will begin to run. Do not mistake this for the metal getting to welding heat. As the heat moves into the pale yellow stage, you'll

Figure 10.3. Set the weld.

notice that the steel takes on a waxy appearance. You are getting very close to welding heat at this point! You will also begin to see a few incandescent sparks coming from the fire, and the metal will be between a pale yellow and an almost white heat, depending on the metal composition.

You are now at welding heat! Quickly remove the link from the fire, lay it on the anvil, and, with a light hammer (1 pound for this particular weld), give a light blow to "set" the weld. Then quickly tap several times on each side to dress the weld and finish it. What you must have at this point is speed and accuracy, not power. There is a difference.

I strongly recommend using a small, lightweight hammer on small welds when first starting out. Too many times, especially when novices begin to do forge welding, they seem to think the harder they hit the iron at welding heat, the better their chance for a successful weld. If your metal is not fluxed properly, the fire is dirty, or welding heat is not reached, swinging a 20-pound sledge will not make the weld take. By using a lightweight hammer, you are minimizing the use of excessive power while increasing speed and accuracy. The other problem in using a heavy hammer is that when the metal is at welding heat, a heavy hammer blow will distort the metal at the point of the weld.

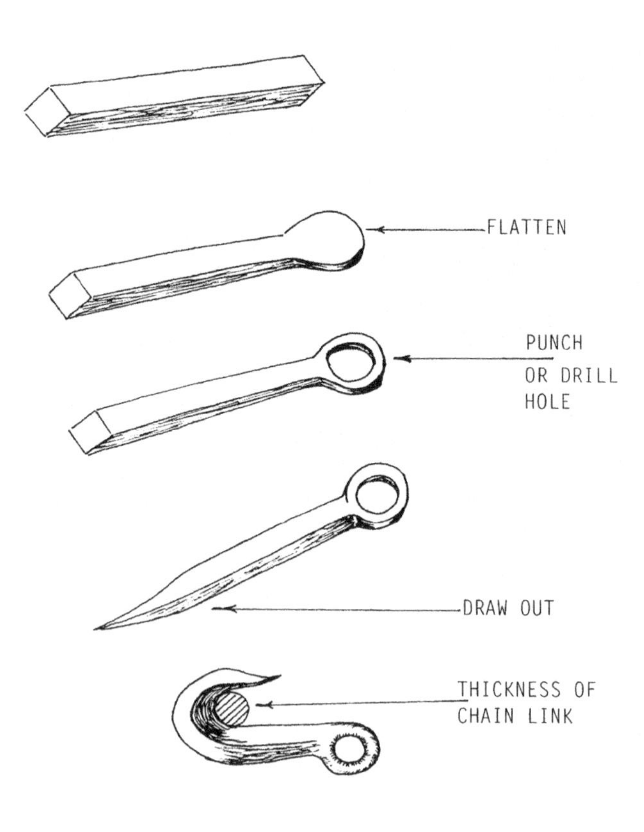

Figure 10.4. Grab hook.

At any time and regardless of the nature of the weld, if it appears that the weld did not take, do not hesitate to reflux and take another welding heat. Better to weld twice than break once.

This, then, is the basic procedure for one of the more interesting and perhaps frustrating aspects of blacksmithing. It has innumerable variations, and its application is limited only by your imagination. For thousands of years, this was the only means of welding. It is a tried and proven method, but practice is necessary to become proficient. Most of the following projects will involve some type of forge welding. Beware the frustration factor!

CHAIN LINK HOOKS

Once you have completed the chain, you now need the hooks. There are two types normally used: the grab hook and the more conventional open hook. The two types of hooks will be made like the crowbar for added strength. Both of these hooks are for a 3/8-inch chain link.

Grab Hooks

Stock: 1/2 inch x 2 feet, square

Heat one end (remember the 1/4 turn) and flatten the end to about 3/8-inch thickness. You can make the hole one of two ways. Either punch the hole with a round punch or, using a chisel, cut a slit, and then take a drift pin to open up the hole to the proper size. In either case, be sure to make the hole slightly oversized so the hook can move freely once attached to the completed chain.

Once the eye is completed, measure back from the base of the eye about 6 inches and cut. Heat and draw the end out to a one-sided point and then bend it out slightly. (See Figure 10.4.) Reheat the middle and make a tight bend using a 3/8-inch bar for the jig or whatever the thickness of your links are, for example, 1/2 inch, 5/8 inch, etc. This needs to be fairly precise, since the hook will be used to grab a link in the chain when hooked up.

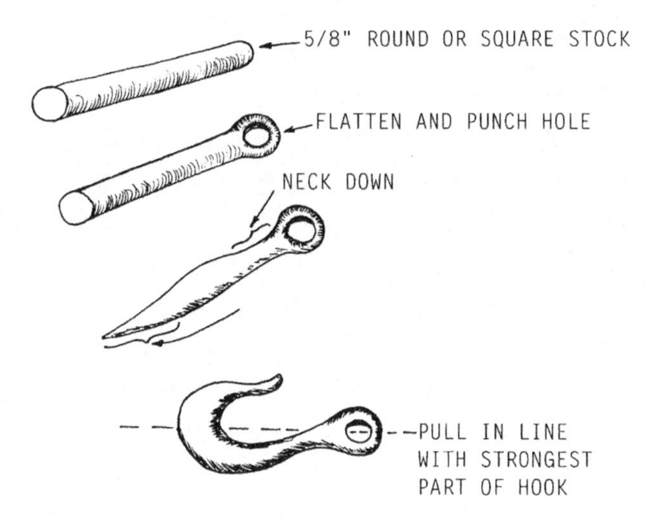

Figure 10.5. Open hook.

Open Hooks

The open hook is made in the same way, except the hook has a larger radius at the bend, resulting in a more open hook, and the body of the hook is drawn out at the radius to give the body more strength at that point. (See Figure 10.5.)

Clevis

Stock: 1/2 x 36 inch, round

The clevis is often used in conjunction with chains as it is usually hooked to the tractor, truck, or team for pulling. Its advantage is that it may be removed after use, and it is easy to slip a chain through when hooking up. In its simplest form, it is a length of rod with a hole punched in each end and then bent in a U shape. (See Figure 10.6) A hitch pin holds it in place.

Heat one end and flatten to approximately 3/8 inch thick. Now, reheat and punch or split a hole big enough to accept the hitch pin. Measure down 10 or 12 inches and cut. Repeat the hole-making process. Heat the center and bend to a U shape. You may want to bring the ends a bit closer to fit over the draw bar, bumper, or whatever, to lessen the chance of bending the hitch pin.

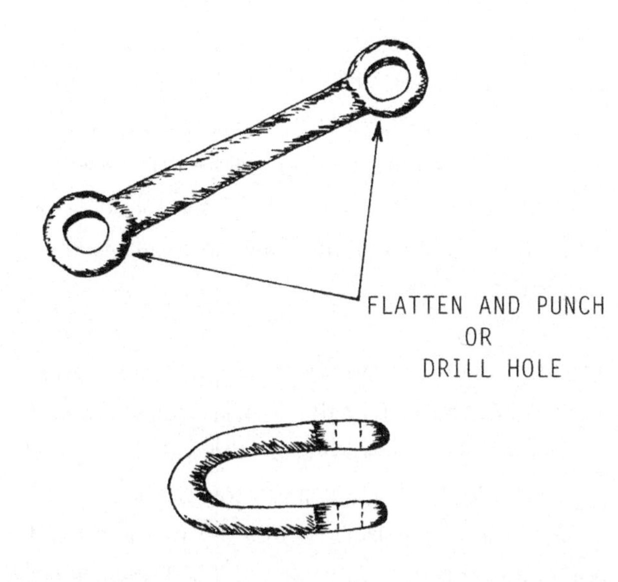

Figure 10.6. Clevis.

TEMPERING

The tempering of steel probably evokes more mystery and is surrounded by more hocus-pocus than any other facet of blacksmithing. We hear of mysterious tempering solutions and elaborate procedures performed in the dark of night by secretive smiths.

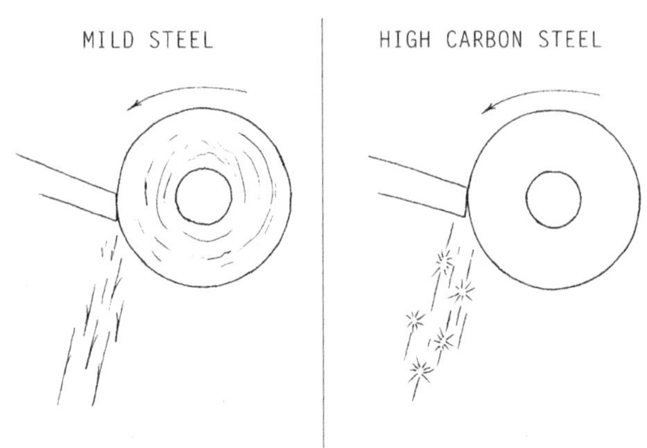

Figure 10.7. Use the spark test to determine the type of steel.

The mystery of tempering is not so much the "secret" solutions and methods as it is the perpetuating of the "secrets" to instill awe in the uninitiated.

Tempering is a three-step process: annealing, hardening, and tempering. Note that tempering and hardening are two different procedures. Many people tend to confuse the two. However, before you run off to temper your knives and axes, keep in mind that we are dealing with nothing but variables. That's where the tricks come in. Two plus two does not always equal four in the mathematics of tempering. The variables are heat, steel, and the quenching or cooling medium. Perhaps the greatest variable is the steel itself, for this will determine the heat required as well as the quenching medium necessary for a serviceable cutting tool or spring. In this day and age of exotic alloys, it has become increasingly difficult to find good old-fashioned carbon steel. This is where the scrap and junkyards can be invaluable.

Let's begin with the steels. High-carbon steel is iron metal of 0.3 to 2.2 percent carbon. Less than 0.3 percent carbon is untemperable "black iron" and mild steel. More than 2.2 percent carbon, as in cast iron, renders the material too brittle for forging. Perhaps the easiest method of finding out if your material is high-carbon steel or mild steel is the spark test. (See Figure 10.7.) Another method is that of "ringing" the piece on the anvil. With practice, you can ring a piece of steel on the anvil and tell if it is high-carbon or mild steel. High-carbon steel will have an almost musical ring to it as opposed to the dull ring of mild steel.

If you are buying your steel from a supplier that specializes in high-carbon steels, you will get a metallurgical analysis of the steel, which will tell you how it should be tempered. The quenching medium of other steels, such as 0-1 and W-2 (oil quench and water quench, respectively), is denoted by their prefixes. It should be noted at this point that for the most part, spring steel and metal files are generally an oil-quench high-carbon steel. Horseshoe rasps, wood rasps, and magneto shafts are water-quench high-carbon steels.

This brings us to the quenching medium itself. Here is where much of the "magic" is perpetuated. Hardening and tempering is the relative speed at which steel is cooled from the outside to the center. This is accomplished by the medium used. For our purposes, cold water is the fastest, while oil, lard, or animal fats are the slowest, excluding air cooling. If your steel is an oil-quench steel and water is used as your quenching medium, "checking" or cracks will appear in the piece. If the steel is water quench and an oil bath is used, the steel may not be hard enough to hold a cutting edge. You can begin to appreciate the variables.

Now we have to contend with the heat. If the steel is heated too hot, it may become too brittle to be of any use no matter which medium is used. If heated too low, the opposite may occur. Either way, it will not perform its intended task. To bring all of these variables into a working order requires practice.

To determine the quenching medium and heat, begin by taking a piece of unknown stock, heating it to a red heat, and pounding it out to a long, flat taper. Now reheat to a good even red heat and bury it in a pile of ashes until cooled off. This is called annealing. On heavy or thicker pieces such as axes, pickaxes, large chisels, and knives, I normally do not anneal them. Smaller knives with thin blades are another matter and should be annealed. It allows the molecular lattice structure of the steel to realign itself and thus relieves the stress created by forging that may cause warpage or cracks when hardening.

After the piece has cooled, which may take several hours, reheat to a good yellow orange at the tip. It will be darker red as it moves back from the tip. (Figure 10.8.) Then quench in water. If it cracks or checks, you now know that you probably have an oil-quench steel. Repeat the entire process, only now quench it in oil. This is the hardening process referred to earlier. At this point the steel is very hard and brittle, totally worthless as a tool until the temper is drawn.

Remembering where the yellow orange, orange, dark orange, red, and dark red lines were prior to quenching, begin to break these points off over the edge of the anvil. At the very tip, you will probably get a very coarse grain structure, almost like cast iron. As you advance toward the red color, you will notice a progressively finer grain until you reach a very tight grain structure, grayish in appearance. This is the heat at which you will quench this particular steel. Normally a good even red heat is best, but it's still a good idea to check it.

At this point, for this particular type of steel, you have eliminated virtually all of the variables. You have established that it is high-carbon steel with the spark test, what the quenching medium is (oil or

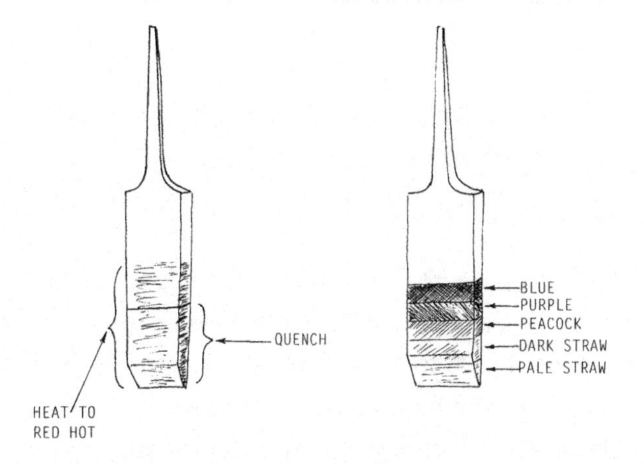

Figure 10.8. Wood chisel.

Figure 10.9. The color spectrum for tempering.

water), and the quenching heat required (by breaking it over the edge of the anvil). All that remains is to form the steel into a tool.

Let's suppose that we have established that this piece is a water-quench steel at even red heat, and we decide to make a woodcarving chisel. First, form the tang by drawing out. Now, move down the blade about 4 inches and cut off any excess steel. Heating the tip of the broad end, a bevel may be rough forged on the end. Heat the cutting edge to a red heat and bury it in the ashes to cool slowly (annealing). Once the piece is annealed, reheat to an even red heat and quench in water part of the way up the blade until the lower half is cooled. Remove from the water and quickly polish the lower half with a sharpening stone. At this point you will notice the color spectrum begin to appear at the end of the polished blade. (See Figure 10.9.) It will start with a pale straw color (very hard) and then proceed to a dark straw (softer), pale blue (softer), peacock blue (softer), and purple (softest).

When the dark straw reaches the end or cutting edge of the chisel, quench the entire piece in the water until cooled. This is the basic tempering procedure. Now, if you have a broad chisel or a large cutting tool like an axe, and you see that one side is "running out" faster than the other after it has been quenched the first time, wait until the one side reaches the desired color at the edge and quench just that portion to "slow it up." When the rest of the color spectrum reaches the right color along the entire edge, quickly quench the whole piece.

When grinding the finished edge on the chisel, be careful not to overheat the cutting edge. If it turns blue from overheating, you have lost the temper and the entire tempering process must be repeated.

CHISELS

There are three types of chisels: cold chisels, tang chisels, and socket handle chisels. The cold chisel is used for cutting mild steel. It is made of one solid piece of high-carbon steel and may have either a double beveled cutting edge or single bevel. The other two chisels are used for carving wood and are normally one-sided bevel chisels.

Cold Chisels

Stock: 3/4 x 6 or 8 inch, round or square (high-carbon steel)

Flatten one end and draw out to the desired thickness. Grind or file the bevel and temper to a dark straw or peacock color. If you do not have a large

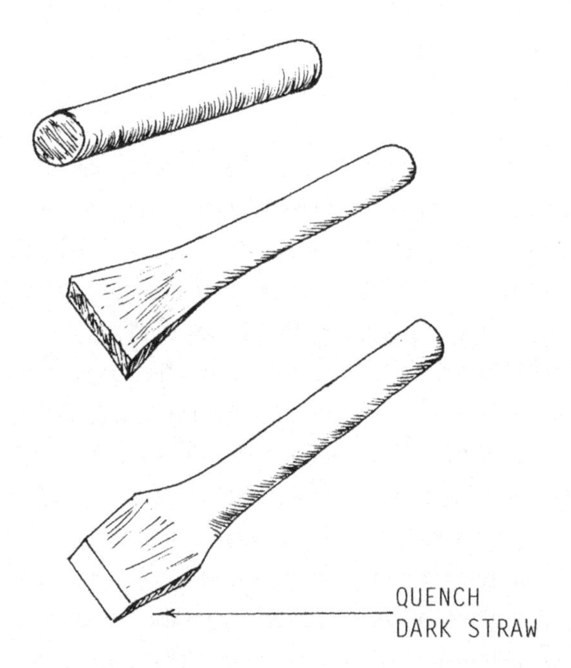

Figure 10.10. Cold chisel.

Figure 10.11. Completed tang chisel.

enough piece of high-carbon steel, take a piece of mild steel (same dimensions), heat one end to a good orange yellow heat, split with a chisel, flux, insert a small high-carbon bit, and forge weld together. (See Figure 10.10.) Now, draw the end out and temper as before. A note of caution here: when quenching a one-piece chisel, be sure once the proper color is achieved and you are ready to quench the last time, there is no red above the color. Should it still be red above the colors and you quench the entire chisel, that red area will be hard and brittle. At some point when you least expect it, the chisel will break when you hit it. You can guess the end result.

Tang Chisels

As the name implies, this type of chisel has a tang running up through the handle like a file. (See Figure 10.11.) In fact, in a pinch, an old file may be used as a chisel. Merely heat and draw the tang out longer, heat and anneal the entire file, file or grind the bevel for the cutting edge, reheat the end, and temper to a peacock or dark straw color. As with the cold chisel, if you do not have a large enough piece of high-carbon steel, take a piece of mild steel $^1/_4$ x 1 x 12 inches. Before forming the tang, take a small piece of high-carbon steel flattened for approximately 1 inch.

Start by heating the end of your mild steel to a good red, pull from the fire, flux, and return to the fire. Next, while the mild steel is still in the fire, heat your high-carbon steel, flux, and return it to the fire. Remember, your high-carbon steel will come to welding heat faster, hence the reason for the sequence of heating and fluxing. If one piece appears to be heating faster than the other, pull that one back from the center of the fire a bit. It is critical that both pieces come to welding heat at the exact same time!

When they are at the proper heat, remove them from the fire, quickly place the chisel stock on the anvil, lay the high-carbon steel on top, and hit it with your hammer. Cut off any excess on the sides. If you have any doubts as to whether you have a good weld, re-flux and return to the fire, bring to welding heat, and finish up on the anvil. Better to take a second welding heat and make sure you have a good weld than to have part of the steel "face" come loose during the tempering process. Now proceed to draw the tang out, shape, bevel, and temper the blade.

For the handle, take a suitable piece of hardwood, drill a hole slightly smaller than the tang lengthways through the entire handle. Rough shape the handle, and then cut from a piece of steel or copper tubing a $^3/_4$-inch piece slightly smaller than your handle stock. This will be the ferrule at the base of your chisel handle. A ferrule is the small metal band used to keep the handle from splitting out either at the base, as in the tang chisel, or at the top, as with the socket handle. Drive the entire handle assembly down over the tang. For the end cap on the handle, cut a piece of $^1/_8$-inch sheet steel the same diameter as the end of the handle, drill, and countersink a hole in the center. Slip this over the tang and peen

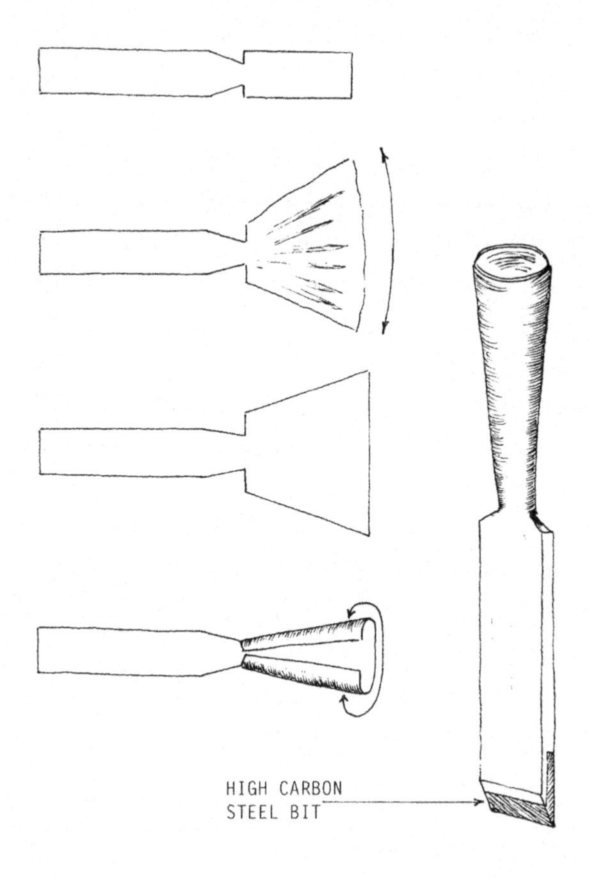

Figure 10.12. Socket handle chisel.

it tight. Finish out the handle to your own preferences. This is one reason many old tools may not feel comfortable to work with. They were made to fit the individual craftsman, not mass-produced to the "one size fits all" concept.

Socket Handle Chisels

Blade preparation is the same as for the tang chisel. Now, however, the handle will be held in place by a socket, much like a shovel. Once you have estab-

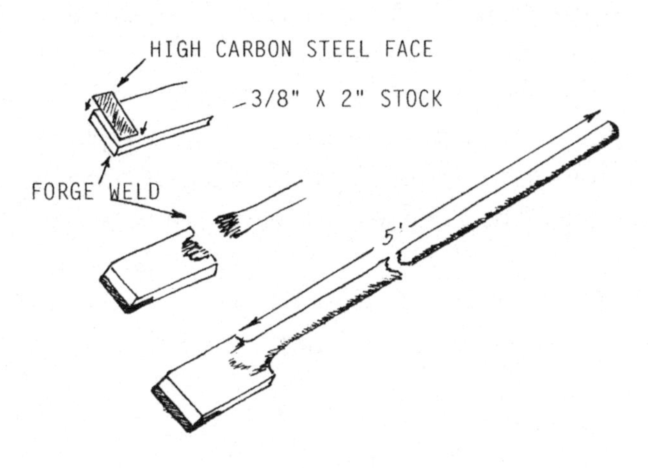

Figure 10.13. Spud bar.

lished the length of your blade, make two cuts, or fuller if you like, leaving approximately 3/8 inch in the middle. Measure up 3 to 3½ inches and cut loose from the parent stock. This 3½-inch portion will become the socket. Heat this area to a good red yellow heat and with the cross peen (not the flat hammer face) begin drawing out as shown in Figure 10.12. You are looking for a triangular shape when finished. Now, true up the two outside edges. This will give you a clean, tight fit after the socket is complete. To roll the socket into the proper shape, heat and begin the roll by lightly hitting the edges equally on each side to start forming the cone shape. This procedure will take several heats. The final truing up may be done on the horn of the anvil or a small bick. Once the socket is complete, take a local heat at its base and tap the base of the socket toward the center and align the socket with the blade.

Taper the handle for a press fit into the socket. You may want to drill and pin the handle. Unlike the tang handle, you will want to put a ferrule at the top of this chisel to keep the wood from splitting or mushrooming out. This type of chisel is normally used for pounding with a hammer or maul.

When shaping any of your chisels to a curve or half-round configuration, shape it as if it were part of a cone.

SPUD OR PRY BARS

Stock: 1 inch x 5 feet, round or square

These are used for everything from chipping ice and prying out boulders to packing dirt around fence posts. Since you want something with some heft or weight to it, I use at least 1-inch-round stock 5 to 6 feet long.

It is advisable to have a high-carbon bit forge welded in, either inserted like the axe or laid on like the chisel, since this tool is often used for chipping ice or breaking up rock. (See Figure 10.13 on page 103.) For a one-sided bevel, the laid-on face is best, as the spud bar is nothing more than a large chisel.

Begin by heating up the very tip to a good yellow heat and upset it. The width and size of the chisel end will determine how much to upset the end. If you wish a wider, heavier blade than is obtainable by upsetting, you can weld a piece on the end using a cleft weld.

CLEFT WELD TO PRY BARS

Stock: 2 x 3/8 x 24 or 25 inches

Forge weld a steel face to one end of the stock as with the chisel. Now measure back about 4 inches from the welded steel face and cut. Scarf the center of the cut end the width of your bar, back about 1 to 1 1/2 inches. Heat one end of the bar and split approximately 1 to 1 1/2 inches deep. Bring both pieces to a good red heat, flux, slip the blade into the open end of the bar, tap lightly to hold in place, flux again, return to the fire, and bring to welding heat. A word of caution here: do not hit the weld hard with the first blow or the blade is liable to go shooting out across the shop. Tap it lightly to seat the weld, and then you can hit it harder to finish it up. You may want to put the blade up against the front edge of the anvil (the base of the horn) when seating the weld.

CENTER PUNCHES

Stock: 5/8 x 12 inch, round or square (high-carbon steel)

One of the handiest tools is the center punch, which you will be using to mark your steel for drilling, heating, cutting, and working the steel in a specific area. Chalk marks can be lost during heating whereas the center punch mark will stay put.

I prefer water-quench steel for the center punch. Heat the end of the stock to a good red heat and draw out to a diameter thickness of approximately 3/8 inch. Now, grind or file the tip to a square point and temper to a medium straw. Cut to

Figure 10.14. A small claw hammer.

approximately 3 to 4 inches in length and bevel the end. The reason for a squared taper for the point is so that it will show up better as a mark at a red or higher heat.

HAMMERS

The method described here for the claw hammer or carpenter hammer may be used for any type of hammer, from the small engraving hammers to the larger blacksmith hammers and sledges. There are two ways to form the eye. One is using a chisel and mandrel, as in the making of the belt axe (see page 112). The other method is to use a punch the size of the eye and punch the eye out. I have found the chisel-and-mandrel method seems to work best for the smaller hammers while the punched-eye method works well for the larger hammers.

Claw Hammers

Stock: 3/4 x 1 x 24 inch

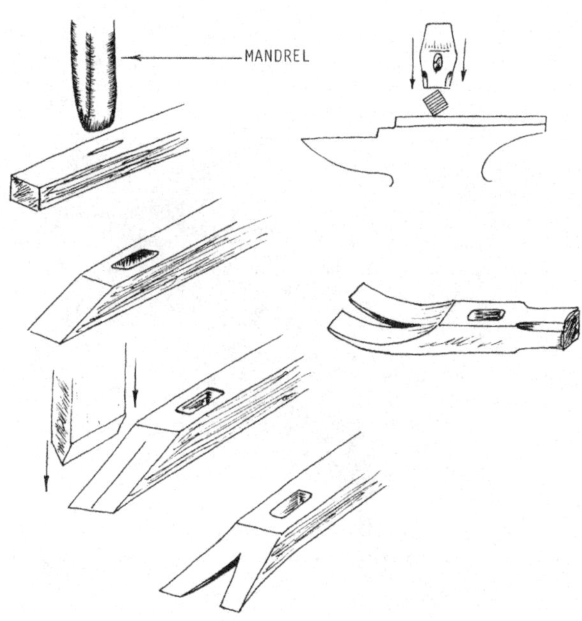

Figure 10.15. Claw hammer.

Begin by measuring in 1 1/2 inches from the end and, using your center punch, mark where the eye will be. Do this on both sides. Proceed as with the small belt axe to make and shape the eye and ears (optional, see pages 116–117). Heat the end of the stock, and moving about 3/8 to 1/4 inch up from the eye, place on the end of the anvil and draw out to a flat taper. This stock will be the claw for pulling nails. Now, reheat to a good reddish yellow and begin to split the claw with a wide beveled chisel. Make this cut as smooth as possible. (See Figure 10.15.)

Once the split is complete, measure from the *other* end of the eye, approximately 2 to 2 1/2 inches, and cut off. This will be the hammer face. You may

Figure 10.16. Small cross-peen hammer.

Figure 10.17. Punch the eye.

want to bevel, flare, or fuller this portion of the head according to your personal taste. Grind or file to a slightly convex or rounded face and bevel the edges. Now, reheat the claw portion of the hammer, spread slightly, and, working over the horn of the anvil, bend the claw downward in an arc. Your hammer is ready for the handle.

Cross-Peen Hammers

Stock: 1 x 1 x 24 inch

As with the claw hammer, measure in about 1½ inches and mark one end of the eye on both sides. Now, instead of using a chisel, you will be using a punch the shape and size you want the eye to be. I use a ½ inch x 1 inch rounded-edge mandrel. Heat to a good yellow heat and punch most of the

Figure 10.18. Draw out and shape the hammer peen.

Figure 10.19. High-carbon-steel face prior to welding.

Figure 10.20. Flux the hammer.

way through. (Figure 10.17.) This will take several heats. Be sure to keep the punch cooled so it does not mushroom. It also helps, once the hole is begun, to drop a pinch of cinders in the hole after it comes out of the fire. This will protect the tip of the punch. Once you can see a dark spot on the opposite side of the hole, roll the piece over and punch out the plug by punching through the dark spot.

Now that the eye is punched, heat the end and draw down to a blunt taper perpendicular to the axis of the eye. (Figure 10.18.) (A straight peen would be parallel to the axis.) Measure 2 to 2½ inches in front of the eye and cut off with a chisel, hacksaw, or hardy. Bevel the sides and dress the face as with the claw hammer.

Figure 10.21. Flux the hammer with the high-carbon-steel face in place.

High-Carbon Steel Hammer Face

To give any pounding surface more durability and toughness, a high-carbon steel face may be forge welded to the hammer face as follows. You may want to do this weld before you start making the rest of the hammer. You will have the parent stock to hold onto, which will make it easier to do this particular weld.

First, shape the high-carbon steel to the same dimensions as the face of the hammer (in this case 1 x 1 x 1/4 inch thick). (Figure 10.19 on page 107.) When it is shaped, heat to a good red heat and, with the corner of a flat chisel, peel up several (4 to 6) "hooks." Let the high-carbon piece cool completely. Heat the hammer face to a good red heat and flux. (Figure 10.20 on page 107.) Take the high-carbon face with the hooks on the back and gently tap this onto the red-hot face. The hooks, or projections, will dig into the softer red-hot steel and hold the face in place. Reheat a bit and remove from the fire just long enough to flux the entire end, high-carbon steel and all. (Figure 10.21 on page 107.) Return to the fire, bring to welding heat, and then carefully slide the piece from the fire and hit it on the end several times to set the weld. Flux the entire end once more, return to the fire, bring to welding heat again, pound the end, and finish shaping on the anvil. Once again, finish to a slightly convex shape and bevel the edges.

You are now ready to temper the face. Here we're looking for the peacock/dark straw blue. If you temper too hard, pale straw or lighter, for example, the surface will likely chip when hitting something hard. You want toughness, not edge-holding ability.

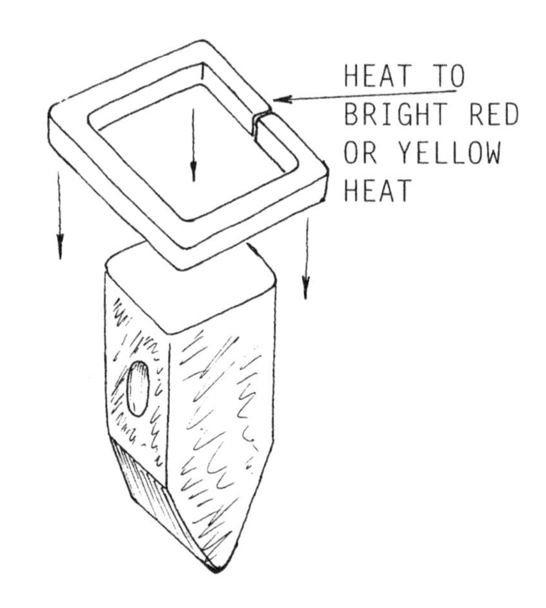

Figure 10.22. Tempering the steel hammer face using a heated collar.

Polish the hammer out. Now, heat the face, at least back to the eye, to a good red heat. Quench fairly deep, but be sure to leave enough heat at the back to "draw" the temper. When the peacock color reaches the high-carbon face, quench the entire piece until cool.

Another method for tempering the face is as follows: Heat and harden as with the first method, only quench the entire head. Now, depending on the size of the hammer, make a band out of suitable stock, say 1/2 inch square, that will fit around the edge of the face. Heat the ring to a yellow heat and slip down around the edge of the face.

(See Figure 10.22.) The color spectrum will move in from the edge. When the dark peacock or straw color is reached in the center, quench and knock the ring off. The advantage of this method is the edge of the hammer face will be softer than the rest of the face and will not chip. The steel face was added to the polled axes as well as hammers.

Do not get discouraged if it takes several tries for this weld. It is one of the trickier ones. The biggest problem is making sure the hooks or projections are long enough to embed themselves in the red-hot face. You must also watch your heat *very closely,* as the two pieces are together (not separate as the laid-on steel) and the high-carbon steel cannot be pulled back to let the other piece catch up, nor is it protected as with the axes where there is a piece of mild steel on either side. So watch your heat very carefully.

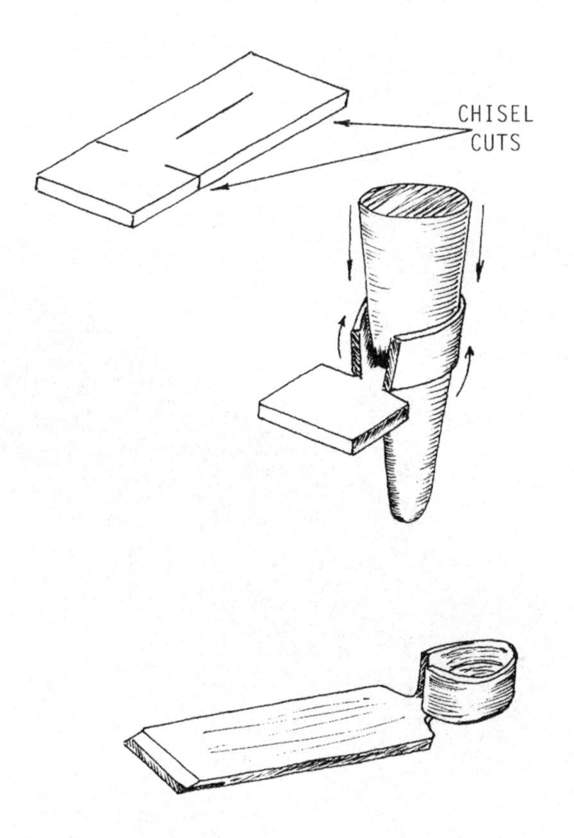

Figure 10.23. Adze.

ADZES

Stock: 1/4 or 3/8 x 2 x 24 inch

The adze is one of the most useful tools for woodworking ever made. As with any tool, it takes practice to become proficient in using it.

Begin by forge welding on the high-carbon steel cutting edge as was done with the chisel. (See page 101.) From the end, come back 3 to 4 inches, depending on how long you want the blade, and make a single cut or narrow fuller on each side, leaving approximately 3/4 inch in the center. From the cuts, move up several inches and, with a chisel, split through the center of the 2-inch-wide stock. (See Figure 10.23.) Now, take a mandrel and drive it through the slit. You will notice how the sides fold up as the mandrel is driven in. Start with a tapered *round* punch (mandrel). You can reshape the eye later. Once the eye is trued up, you may want to put a slight curve in the blade of the adze, but this will be determined by your preferences and how you will be using the adze. At this point, you can shape the blade to your particular needs—half-round, flat, slightly curved, whatever you want. In effect, all of your tools are custom

Figure 10.24. Spanish adze with interchangeable blades.

made. Try getting *that* at a local hardware store or discount house.

Spanish Adzes

Another type of adze that is rather unique and found in the Southwest and Mexico is commonly referred to as the Spanish adze. What makes this adze unique is the interchangeable blades, so it may be used in a number of different applications. (see Figure 10.24.) The Spanish adze is composed of four different parts: the handle, the iron collar, the wedge, and the blade. The handle should be made of some type of hardwood, and the shape of the blade is only limited by your imagination and the uses of the adze.

The Spanish adze is shown here as an example of another variation of this versatile tool.

AXES

The axe is one of the handiest tools found around any homestead. Over the centuries, it has taken many and varied forms, from the elaborate battle axes of the Middle Ages to the utilitarian woods-

Figure 10.25. Three types of axes are made using different methods of construction. From left, felling, wrapped-eye, and punched-eye axes.

man axe carried by the early pioneers. Next to the knife, the axe is probably one of man's oldest tools.

Construction of the axe is as varied as the design. Some, like the German Goosewing, may require as many as three or four pieces and welds to complete the axe. Others require only one weld and are constructed of one piece. The same two basic procedures we will cover can be used for a full-size felling axe or a small belt axe or hatchet. One is the "wrapped eye" method, and the other is the "punched eye" method. Both require forge welding.

Mandrels

Stock: 1 1/4 x 24 inch, round

Use a mandrel to form the eye of the axe. The shape of the axe eye is a matter of personal preference, which explains why so many different shapes are found on the old tomahawks and belt axes. The eye may be round, oval, teardrop, square, triangular, or anything in between. There may or may not be a taper to the eye. I have examined axes from all parts of the country in museums and private collections and have seen

Figure 10.26. Tomahawk made using the wrapped-eye method.

Figure 10.27. Various mandrels used in axe making.

both types, with and without the taper. (See Figure 10.27.)

For the teardrop-shaped mandrel, you will have a two-way taper. Begin by heating the end of the stock and drawing down to a gradual flat taper. Measure up approximately 4½ inches and begin drawing one side out to a rather blunt taper. (See Figure 10.28.) This will give you, in cross-section, the rough teardrop shape. It will take several heats and light hammer work to true the mandrel up. Finish by using a grinder or file for final shaping. Cut from the parent stock and bevel the top.

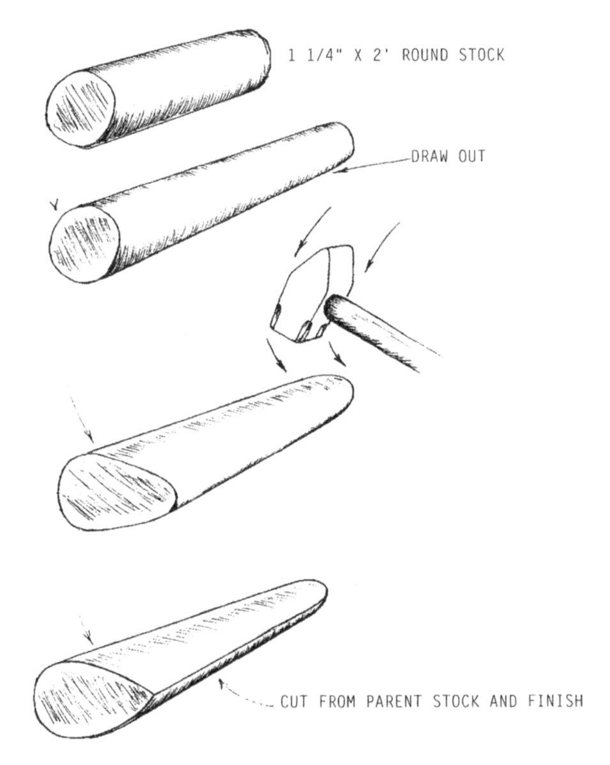

Figure 10.28. Shaping the mandrel.

Wrapped-Eye Axes

Stock: 3/8 x 1 1/2 inch x 3 feet (mild steel)

This method seems to find its most prolific application in the manufacture of the small hand axe or tomahawk and has been used for thousands of years. It is also the first type of axe I learned to make. Billy said he and his father used to make these axes for the woodcutters and charcoal makers in Germany. This axe has no poll and is usually relatively lightweight, weighing 1 to 2 pounds. This is not to say there weren't axes of larger sizes made in this manner.

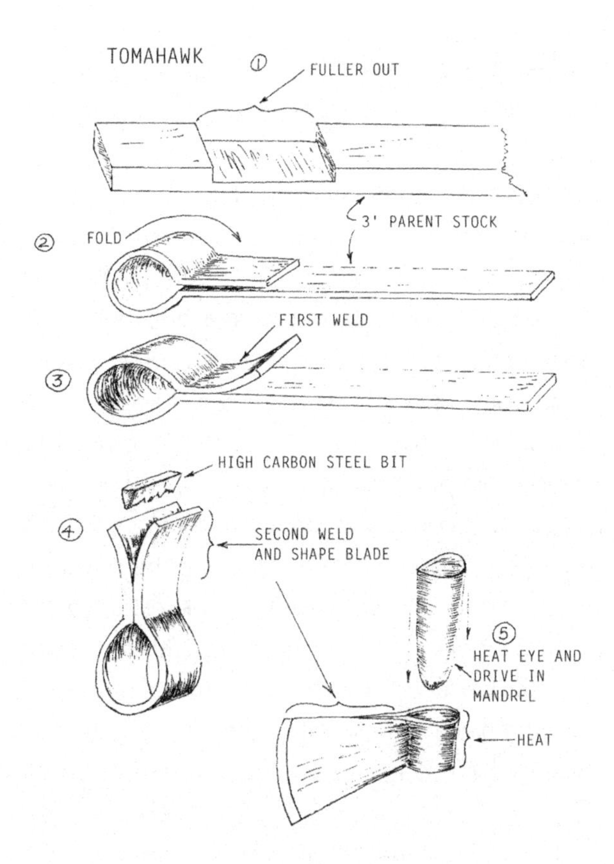

Figure 10.29. The wrapped-eye method.

It is also the method used for making tomahawks or trade axes bartered to the American Indians during the 1700s and 1800s.

You will need a piece of high-carbon steel 1/4 x 3/8 x 1 1/2 inch for the cutting edge. To make this piece, merely heat and draw one side down lengthwise. It will look like a 1 1/2-inch-long triangular-shaped piece. After heating the high-carbon bit to a red heat, take a set chisel or, on the hardy, rough up the sharp edge of the bit. Set aside and let cool.

On the 3/8 x 1 1/2-inch stock, measure back 2 inches from the end and make two center punch marks on both sides of the stock. Now take your mandrel (teardrop shape works best), lay one edge on the two center punch marks and roll it 180 degrees, add a 3/16-inch mark, and center punch both sides. This area will be drawn to approximately 1/4-inch thickness and will become the eye of the axe. Use the cross peen for the initial roughing out, and then smooth it out using the hammer face. When finished, you will have a shoulder of sorts on each end. Heat the fullered eye area to a good red heat and roll it so the shoulders meet; flux, tap together, and bring to welding heat (See Figure 10.29.) Where the shoulders meet will be your first weld. Heat to a good red and flux *only* the area around the shoulder. Bring the area around the shoulder to welding heat. When setting the weld, I've found it works best to do it on the edge of the anvil.

Figure 10.30. Bevels for axes.

Since there is going to be a tremendous amount of strain at this spot, I usually flux and take a second welding heat. Better that than having the weld come apart when you are shaping the eye. After this weld, take your mandrel and pound it into the bottom of the eye about halfway. Return the axe to the fire and bring the back of the eye to a good yellow heat and drive the mandrel in from the top far enough to shape the eye. On this type of axe (tomahawk), I make the eye wider at the top. The handle is put in from the top, like a pickaxe, so no wedging is necessary.

Reheat the end of the overlap (the edge end) to a good yellow heat. Using the hot cutter or chisel, cut it off flush, and then pry it open slightly using the hardy. Quickly flux the end, insert the high-carbon bit, pound together, re-flux, return to the fire, and bring to welding heat. The shaping of the blade as well as the weld itself can be accomplished with this one heat. Additional low heats may be required for finish work and trueing the eye, that is, aligning the eye parallel to the mandrel.

With the axe completed, all that remains to be done is tempering the cutting edge. For axes it is best to quench at a dark straw color. After the tempering process, take the axe and grind the bevel. Be careful not to overheat it on the grinder. If you grind until it turns blue, you have lost the temper and must repeat the tempering process. A good rule of thumb is to have it spend as much time in the coolant as on the grinding wheel. It should be noted that when grinding the bevel on any axe, it should be a rounded bevel, not a flat bevel (See Figure 10.30.) The rounded bevel not only cuts better, it facilitates the removal of the axe from the piece of wood after the cut. For a throwing axe, since you want it to stick, use a flat bevel.

At this point it might be worth mentioning the practice of "wedging" the handle. If you have no taper to the eye of an axe or tomahawk, a wedge in the end of the handle will be required to secure the head and handle. When placing the wedge, never run it parallel to the top of the blade. This puts tremendous strain on the base of the eye. In a wrapped and welded axe as we just described, this extra strain could result in cracking or splitting the weld. It is best to place the wedge at an angle or even perpendicular to the plane of the blade of the axe.

Punched-Eye Axes

Stock: at least 3/4 x 1 x 24 inch, (mild steel)

The punched eye method is a very strong design and is the way in which axes are made today. (Figure 10.31.) illustrates the construction process.

This method works well for small belt axes or hatchets. You will need these tools to make the axe: a 3/4-inch-wide chisel, a long punch, and a mandrel for the eye.

Decided on the size of the blade, usually 2 to 2 1/2 inches, then take a center punch and mark the eye with two marks on both sides. This is where you will begin splitting with the chisel. Reheat to a good yellow heat and split the end length-wise. Flux, insert the high-carbon steel bit, pound closed, return to the fire, bring to welding heat, and weld together. Be careful not to pound the weld too hard at this point and flatten it out. Hit it just hard enough to set the weld. You may also upset the end with this heat if you want a wider blade on your axe. I will normally put it in the vise and pound the end to upset it. Do not shape the blade with this heat as with the tomahawk.

Figure 10.31. Belt axe made using the punched-eye method.

Your next step will be to punch the eye with a chisel. This will take several heats and must be done from both sides. Be sure to keep the cutting edge of the chisel cool by having a can of water near the anvil to dip the chisel in. A word of caution here: do not have the edge too thin on the chisel as it will deform from the heat of the steel. You will also find that a chisel with a somewhat rounded point (see Figure 10.33 on page 116) will cut faster and not be as prone to distortion by the heat.

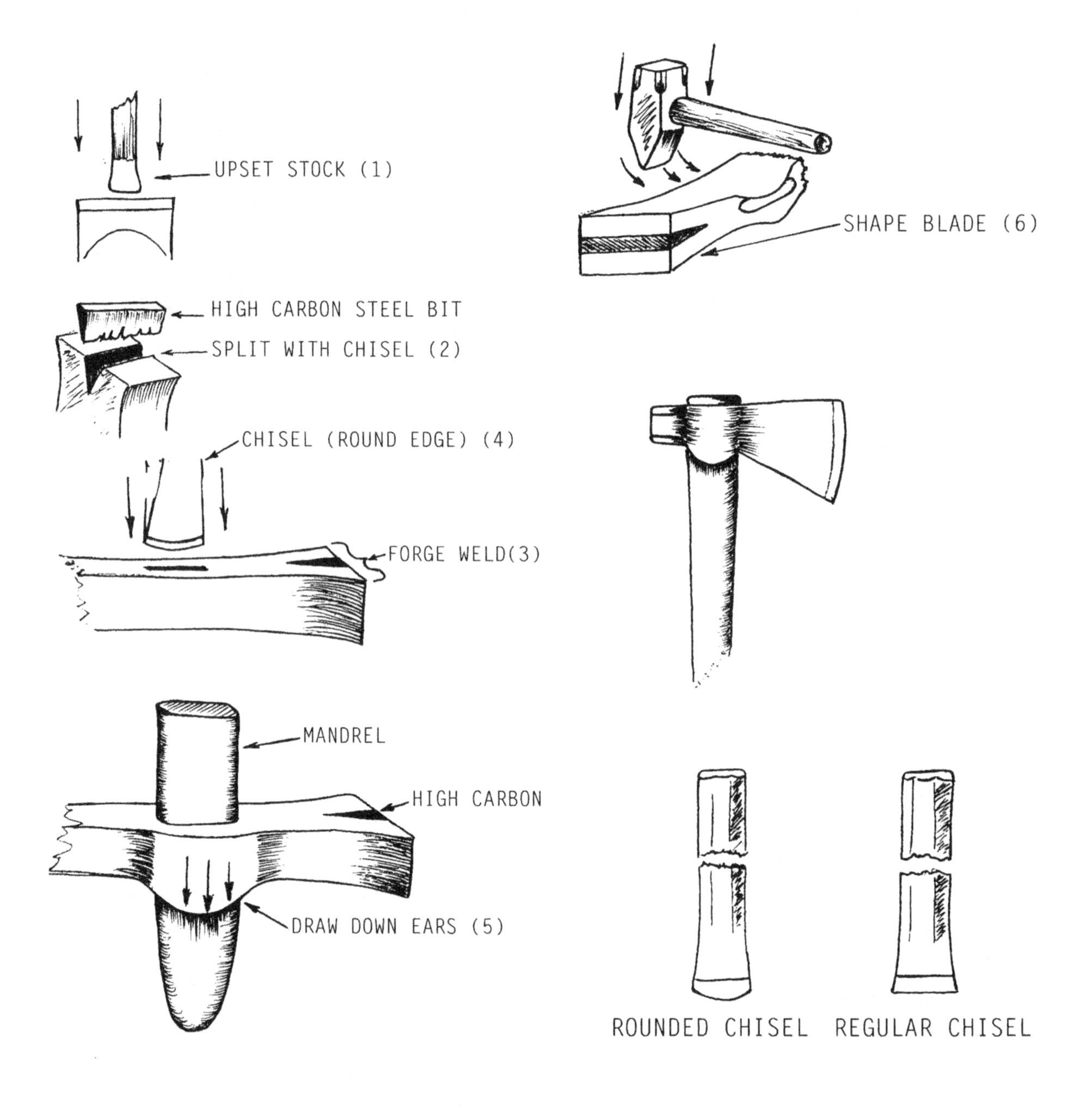

Figure 10.32. Punched-eye axe.

Figure 10.33. Rounded chisel.

Heat the area where you have made the center punch marks to a good red to red yellow heat. Clamp the stock in the holdfast and start splitting with your chisel. After one or two blows with the hammer, remove the chisel and cool it off in the can of water you have near the anvil. When you get about halfway through, turn the piece over and cut a deep chisel mark on the other side before returning the stock to the fire. By doing this, you have a very easy mark to locate the start of the second cut. On 1-inch stock, with a little practice, you can usually cut completely through in two heats.

Once you have cut through, reheat and, using a long punch, (this is a tapered punch, the same shape as the eye only a bit narrower and longer, approximately 12 inches) drive it through the slit to open up the eye. If you wish to forge "ears"—the protrusions that extend below the eye on either side—on your axe, reheat the eye area once again, drive the long punch through the eye, and begin drawing the ears down. Unless you have a trip hammer, this operation will take several heats.

Now it is time to shape the blade of the axe. Heat the portion of the stock *in front* of the eye to a good red or yellow red heat and begin drawing the blade out using the cross peen of your hammer. Again, if you do not have a trip hammer, this shaping will take several heats.

Once you have roughed out the blade shape with the cross peen, use the flat face of the hammer to finish and smooth the blade.

Now that the blade is shaped, heat the eye area to a good red heat and drive your mandrel through from the bottom. Reheat and drive it through from the top. At this time you will want to make sure the blade is vertically aligned with the mandrel (handle). (See Figure 10.32.)

Determine how long a poll you want on your axe and cut it off. Temper the blade to a dark straw and sharpen. A note on sharpening: I normally will start the bevel before I temper the blade. By doing that, I don't have to spend much time finishing out the edge on the grinder and thereby lessen the chance of getting the cutting edge too hot. Remember, if you are sharpening on a grinding wheel and the edge turns blue, the temper is gone and it must be re-tempered.

Felling Axes (Long Polled)

Stock: 3/8 or 1/2 x 3 x 24 inch

Since this is a larger type of axe, there are several ways of making it. The one shown here is the procedure I use. Don't be afraid to experiment. The mandrel shape for this axe should be long and narrow. Once you've made your mandrel, begin by fullering down the eye as shown in Figure 10.34 on page 118. When you have shouldered the eye, turn the stock over and, using a chisel, cut almost all the way through the stock. Flux, fold together, reheat, and forge weld. You may, as in the tomahawk, want to take a second weld.

Reheat the end of the axe, cut off, spread the end, flux, insert the high-carbon-steel bit, flux again, reheat,

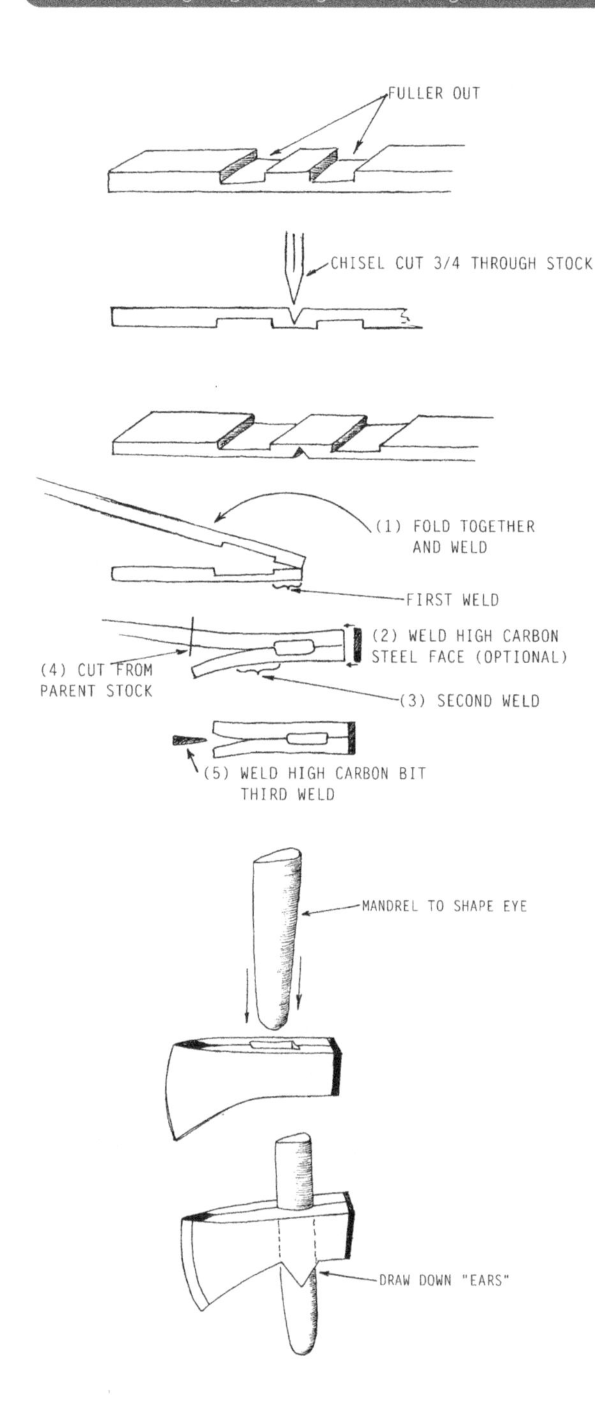

Figure 10.34. Long-polled felling axe.

Figure 10.35. Short-polled felling axe.

and forge weld the blade of the axe. You may be able to weld and shape the blade in one heat if you're making a small axe, 1 1/2 or 2 inches wide, but for a larger axe—3 or 3 1/2 inches wide—you may need two welds: one at the base of the eye as in the tomahawk and the second weld for the high-carbon bit.

Once all welding and shaping is done, reheat, insert the mandrel one last time to shape the eye, and align the blade with the mandrel. Temper to a dark straw.

Felling Axes (Short Polled)

There is another method of axe construction that is not as commonly used. This may be used either in the larger felling axes or the smaller belt axes. The poll may be made a bit larger by forge welding the high-carbon steel face to it.

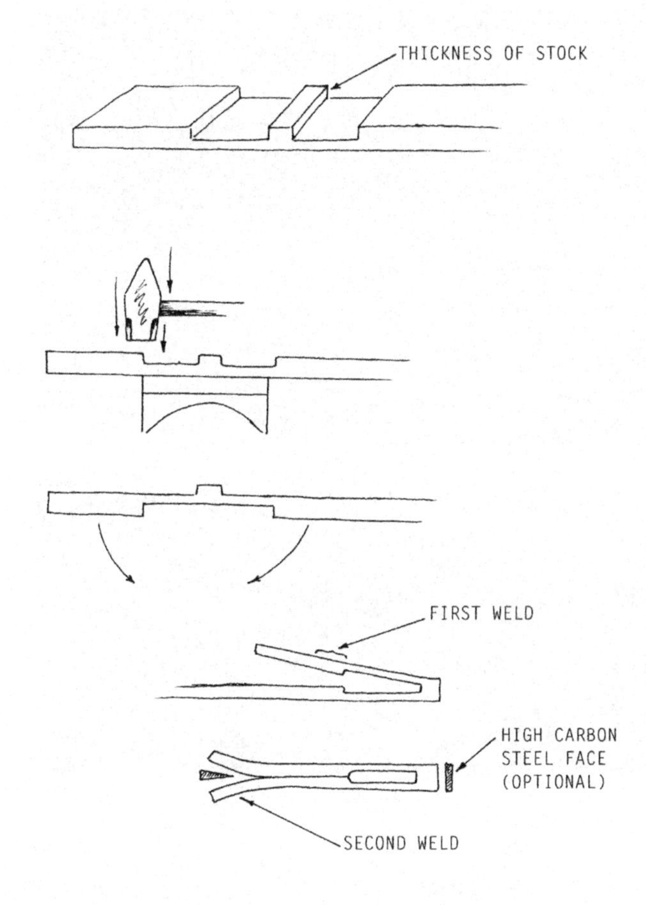

Figure 10.36. Short-polled felling axe.

The process is similar to the long poll axe except that the distance between the two fullered areas is the *thickness of the stock* being used, so for a larger or longer poll, use heavier stock, such as 3/4 x 2 inches or heavier. For smaller axes, 1/2-inch stock works well.

Once the two areas are fullered out, reheat, turn the stock over, and "move" the outside edge to the opposite side as shown in Figure 10.36.

After both edges have been rearranged, heat and fold them together for the first weld. Proceed as shown. Once all the welds are done, you may want to draw the poll out some to give it a bit more length. I have found 1/2-inch stock won't draw out that much, but 3/4 inch will. Once you have drawn out the poll, finish shaping and tempering the blade.

The axe shown in Figure 10.35 was made using this method.

KNIVES

The knife is probably one of man's oldest cutting tools or weapons. It has been in use for thousands of years in various forms. When all other weapons and tools of survival are lost, having a good knife at your side gives one a sense of security.

At some point in time, almost every blacksmith has made a knife. It is one of the first projects usually attempted by the beginning blacksmith, but it is also one of the more difficult. To make one properly requires considerable skill in the areas of forging and heat treating.

Today, as in days past, files seem to be one of the most-used steels for knives. It has been suggested that some of the file teeth were left on many of the early knife blades to assure the customer that he was getting good, high-carbon steel in his knife. Fact or fiction, it is interesting speculation.

Figure 10.37. Three knives based on old designs.

Throughout history the techniques of knife making have been many. Though some are complicated and involved, such as pattern-welded Damascus steel and the Japanese Samurai sword, we will deal strictly with the simpler aspects of knife making. There are two excellent reference books on the primitive knife listed in the resource section. They are very good sources for ideas on knife design.

Many of the early knives appear to favor the file-tang method of construction as opposed to the full-tang, or slab-handle, method. Steel availability may have been a consideration. Since files were used as the steel, and a tang was already partially formed on the file, it is easy to lengthen the tang a little and then continue with the forging of the blade. That is the basic method that will be discussed here.

A word of caution before we begin: Watch your heat! One of the tricks (if there really are any) to successful blade forging is not to have any hot spots and to have even, equal hammer work. High-carbon steel burns easily because of its higher carbon content. Keep your heat at a good orange-red. Do not overheat and get up into the yellow heat, as you'll be getting very near the burning point. If the metal begins to burn, throw the piece out and start over. It is not worth taking a chance on it.

After you have once broken the blade on a finished knife because of improper heating, you'll know what I'm talking about. Since we are using traditional methods, a used 8-inch file will be the stock.

Begin by heating the base of the file tang and drawing the tang out longer to accommodate your hand size. This will be the handle for your tongs while forging the blade. When you begin forging the blade, remember there is a two-way taper involved. It tapers from the back of the blade to the cutting edge and from the base of the blade to the point or tip. I have found that by drawing out the taper from the base to the point first, you will tend to get a bit of a downward curve toward the cutting edge, which is what you want. If nothing else, put a slight downward curve on the blade using the horn of the anvil. Remember, as you draw down the edge of the blade, you are, in effect, stretching that side, which will tend to create an up-turned curvature to the blade. For a straight blade, you will need to start with the downward curve. For a skinning knife, you will want a slight up-sweep to the blade. (See Figure 10.38.) As you start forging the cutting edge, you will notice that the blade will begin to straighten itself out to the right shape, as illustrated. During the blade forging, keep your hammer blows as even as possible and turn the blade every few blows to equalize the forging on both sides of the blade. Also, try to keep the force

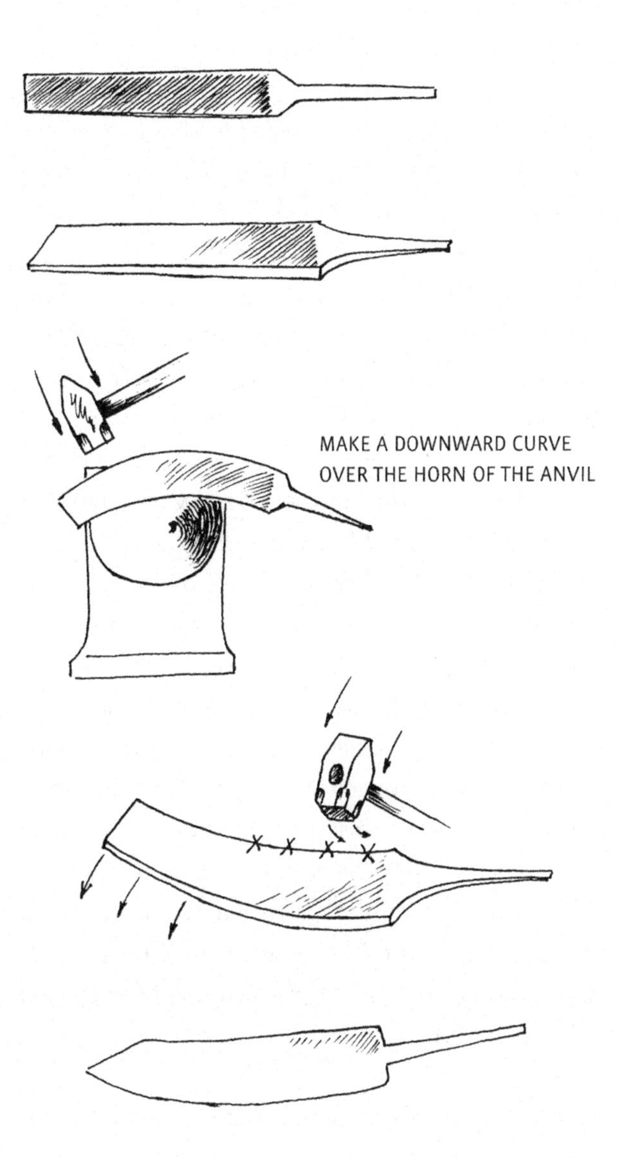

Figure 10.38. Knife blade.

of your blows consistent. If either of these procedures are not followed, warpage of the blade may occur during the tempering process.

Once the blade is rough forged to shape, the next step is "packing." This is done along the cutting edge at a very low, barely discernible red heat and with a light 2-pound hammer. Upon reaching the desired heat, use light, rapid blows on both sides of the blade along the cutting edge. This process may take several heats, depending upon the length of the blade. This packing compresses or packs the molecular grain structure of the steel along the edge of the blade, which makes a tougher, more durable cutting edge. This can also be done on axes. Over the years there has been debate as to whether packing actually improves edge-holding ability. This process was passed on to me by Billy, who said it was used primarily on axes and heavy-bladed cutting tools that were subject to high-impact stress. Keep in mind that, years ago, hand tools such as axes, chisels, and the like were used on a day-to-day basis a heck of a lot more than they are today. I tend to think the techniques used in the making of these tools were based on a lot of practical experience and a lot of hard use.

Following the packing of the blade, it is time for the annealing process. This relieves any internal stresses created during the forging of the blade and is done by cooling the blade slowly, thus allowing the molecules within the steel to realign themselves in their proper lattice structure. In short, it helps to eliminate warpage and stress cracks. It is done by heating the entire knife (tang and blade) to a good cherry-red heat, and then burying the entire knife in ashes or sand. Depending upon the size of the knife, this may take several hours.

After this, the file work and grinding are done to rough out the shape and finish of the blade. Once this rough finish work is done, you can commence the hardening and tempering process. There are several ways to temper and harden. One of the easiest is as follows: Heat the entire blade to a good, even red heat and quench in motor oil (used oil is fine), cutting edge down, parallel to the oil, until cool. If the blade is not warped, proceed to the tempering. If the blade is warped, go back and anneal the blade and repeat the hardening process. One way to straighten a blade is to heat to an even red heat, hold it by the tang with the tongs, and, keeping it vertical with the blade down, hit the tang sharply with a hammer. Then bury it in the ashes or sand to cool slowly (anneal).

Another method of hardening and tempering, and the one I use for knives, is called differential quenching. It has been in use for hundreds—if not thousands—of years and is done as follows:

1) Heat the blade to a good, even red heat.
2) Quench the blade only halfway (Figure 10.39).
3) Hold until the red disappears from the back side of the blade and then immerse the entire blade in the oil until cooled off.
4) Remove the blade from the oil and "flame" the

oil off the cutting edge by passing it over the fire several times, requenching each time to cool.

5) Test the blade.

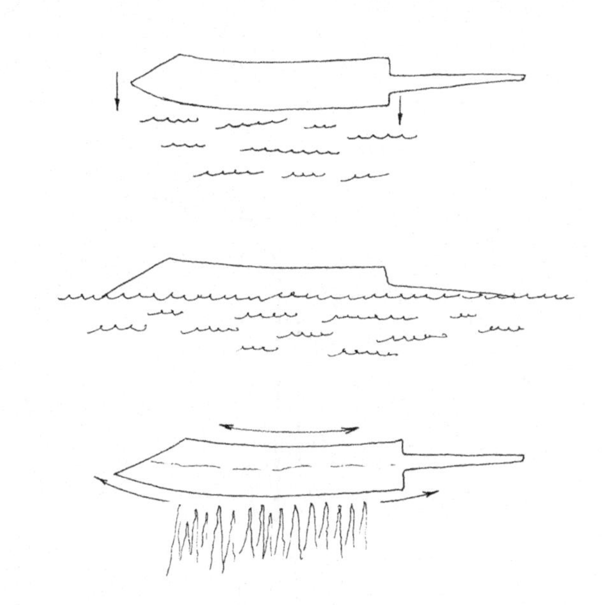

Figure 10.39. Knife tempering.

To test the blade, I use two methods. If tempered properly, a new file will barely cut the blade, so take the file and file the cutting edge as if you were sharpening the blade. This will also reveal any "soft" spots if you didn't get an even heat.

The second method I use is to take the blade, after it's tempered, and chop on a piece of mild steel. If the mild steel is nicked, and there are no chips in the blade, it's good to go. Not very scientific, granted, but it has worked well for me over the years. I have heard of other processes, some of which defy description. Remember, if the temperature spectrum runs out past the peacock blue, you have *no* temper. After 35 years of tempering axes, pickaxes, and knives, I've been down this road. As I said earlier, hardening and tempering are based on basic physical principles, not hocus- -pocus or magic. Keep in mind that the knives of 100 or 200 years ago were used a lot more on an everyday basis than they are today. What worked then will most certainly work in this day and age.

With the blade tempered, work can now proceed on the handle. If you are planning to have a cross guard or quillon, all file work and shaping of the shoulder, etc., should already be done prior to the tempering. You have only to "seat" the cross guard on the shoulder of the knife, and then attach the handle. On many of the early knives, all furniture, i.e. the butt cap and hand guard, was iron. There were several reasons for this: iron is more durable for the rough usage the knife would encounter, and iron furniture can be installed without the use of soldering, brazing, or pinning. One method of installing an iron cross guard is as follows: Take a piece of $^{1}/_{4}$ x $^{1}/_{2}$-inch stock and rough forge the guard as shown in Figure 10.40 on page 124. Drill or punch a hole in the center and, with a small file, true up the eye so that it is a little undersized relative to the size of the knife tang. At this point it is advisable to do the rest of the finish work on the guard. Place your knife blade in a vise with the tang vertical. Take the finished quillon and heat to a good cherry red. Remove from the fire, quickly slip it down over the tang, drive it down on the shoulder of the blade, and cool the whole unit by pouring water on the quillon. When the guard

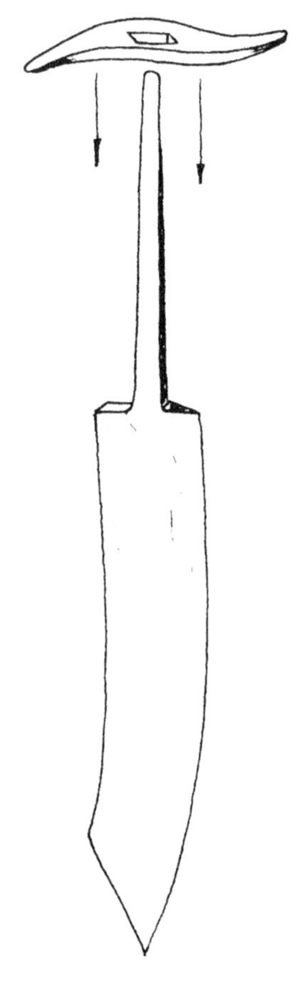

Figure 10.40. Seating the quillon.

is cooled, it will shrink tightly around the base of the tang. This method does not work well if you are using very thin stock for the quillon because there is not enough mass to effectively shrink up on the tang. The rest of the handle may now be assembled, the butt cap fitted, and the end of the tang peened over to hold it all in place. Be sure that the end of the tang is annealed (softened) so that it can be riveted over the end of the butt cap. You now have a rugged, serviceable knife that, if done properly, will last for generations. Perhaps it will end up in someone's collection in the twenty-second century as an artifact.

What has been covered here are merely the basics to what is a most fascinating craft. I have covered very little in the way of ornamental ironwork. That in itself would be another book. With this introduction, you can begin to appreciate the virtually unlimited possibilities. As stated in the beginning, never hesitate to experiment and be willing to pass on information and techniques to others. Blacksmithing is a trade steeped in tradition and, to many, apparent mystery. Perhaps even more mystery is associated with it in the twenty-first century because it does not seem to fit in our computerized world.

11 Finishes and Patinas

Steel, no matter how well finished (especially outdoor pieces like hinges, latches, and door knockers), will eventually rust unless continually maintained. Some pieces are electroplated, painted, or given a rusty antique look by using caustic chemicals. Below are listed some, but by no means all, of the finishes available. Don't hesitate to experiment! For axes and knives, I normally use a light coat of 3-In-One Oil when they are not being used. Cooking utensils should be dried immediately after washing. If being stored for any length of time, apply a light coat of cooking oil.

BRUSHED

This is one of the simplest yet best-looking finishes. Once the piece is done, merely run it over a wire wheel brush. I prefer the coarse "knotted wire" wheel because it knocks off the oxidation scale easier. A softer brush will give a more polished look. This is primarily for smaller pieces, such as those covered in this book. For decorative pieces such as towel bars, curtain rods and finials, and candleholders, the piece can then be sprayed with a clear satin finish lacquer. Do not use the spray on *cooking* or *eating* utensils.

PAINT

Paint may look okay on lawn furniture, but I think it looks cheap on hand-forged ironwork, whether decorative pieces or tools.

OIL/BEESWAX

This works well on smaller pieces such as door knockers, sconces, candleholders, hinges, and other outdoor pieces. After you have brushed the piece, heat to a barely discernable red and brush either motor oil or beeswax on it. Allow to cool, and then rub dry. This will give a uniform, almost black finish. This type of finish can be maintained through periodic application of a light coat of furniture wax.

BRASS/COPPER

Again, smaller pieces work best for this finish. It looks very good when used in coloring leaves or feathers or highlighting an area. Heat the piece to just below a discernable red and, taking a small brass or copper brush, begin to scrub. Once you have reached the desired color, cool, pat dry, and spray with a clear satin lacquer or furniture wax. It may be maintained the same way.

PEWTER

The pewter finish requires more maintenance than the other finishes because steel, like any metal, will oxidize and turn dark when exposed to air. Once the piece is polished, it must be sprayed with clear satin lacquer or furniture wax immediately.

After the piece is cooled, use a high-speed, coarse knotted wire wheel to knock off *all* the scale to give the bright brushed steel (pewter) look. As soon as it is polished completely, seal with the aforementioned coatings.

BLUING/BROWNING

Prepare the piece as with the pewter finish, polishing off all the scale. Then, depending on the color you desire—either blue or brown—apply the proper agent. I have had very good luck with Birchwood Casey's Super Blue and Plumb Brown solutions. These can be found at almost any sporting goods or gun store. Used in combination, they give almost a black finish. Silver Black, found in jewelry supply houses, will also work. Let the solution dry and finish by lightly polishing with 0000 (4 aught) finest steel wool. If the metal is slightly warm, the patina will set better. Maintain with light oil or furniture wax.

POTASSIUM PERMANGANATE

This solution will give an antique appearance to any organic or carbon-containing material, like wood, leather, and steel. Potassium permanganate is an oxidizing agent. It comes in powder form (purple in color) and can be obtained from almost any chemical supply house. Mix with water to the desired strength and apply with cotton swabs. *Caution:* Wear rubber gloves, as this will turn your fingers or any other skin area it touches a dark brown. Not the type of thing you want to have happen the day of a black-tie social event.

I have used this solution to great effect when antiquing polished knife and axe blades. Merely polish the blade, apply as many coats as required to achieve the desired patina, and polish out with 0000 steel wool.

MURIATIC ACID

This is another antiquing agent. Unlike potassium permanganate, this is very caustic and should be used in a well-ventilated area. Polish the piece and immerse it in the acid until the desired finish is achieved. Once this is done, rinse the piece in soda water to neutralize the acid. I have used this solution for my Damascus knives and axes. Muriatic acid (diluted form) may be purchased from most hardware or home improvement stores since it is used in concrete work.

VINEGAR AND BLEACH

Yet another antiquing agent is the homemade variety. Mix equal amounts of vinegar and bleach in a plastic or glass container. Polish the piece out as in the pewter finish and immerse or apply the solution until the desired finish is reached. Remove and rinse in soda water, a neutralizer base, which will stop the chemical reaction. Polish lightly with 0000 steel wool.

CERAMIC GLAZES

Ceramic glazes come in a powder form and can be mixed with clear shellac for subtle coloration and highlighting. Unlike spray paints, they don't give the cheap mass-produced factory look. They can be maintained using furniture wax.

These, then, are just a few of the coloration techniques that may be used to color or antique hand-forged iron. As mentioned in the beginning, do not hesitate to experiment.

Resources

Works Cited

Book of Buckskinning, Vol. IV, Chapter 1. Texarkana, TX: Scurlock Publishing Inc., 1987.

Suggested Suppliers

Abrasives

A Cut Above Distribution Co.
(800) 444-2999
www.acutabove.com

Industrial Abrasives Co.
542 N. 8th St.
PO Box 14955
Reading, PA 19612
(800) 428-2222
Great service and great prices.

J&L Industrial Supply
PO Box 3359
Livonia, MI 48151-9918
(800) 521-9520

Klingspor Abrasives, Inc.
2555 Tate Blvd. SE
Hickory, NC 28603-2367
(800) 645-5555

Sparky Abrasives Co.
6040 Earle Brown Dr.
Brooklyn Center
Minneapolis, MN 55430
(800) 328-4560

Blacksmith

Centaur Forge
117 N. Spring St.
Burlington, WI 53105
(262) 763-9175
(800) 666-9175

Forge & Anvil Metal Studio
30 King Street
St. Jacobs, ON N0B2N0
Canada
(519) 665-3622
www.forgeandanvil.com

Glendale Forge
Monk St.
Thaxted, Essex CM62NR
England
011-44-1371-830466
www.isys.co.uk/glendale

Kayne & Son
100 Daniel Ridge Rd.
Candler, NC 28715-9434
(828) 667-8968
www.kayneandson.com

NIMBA Anvils
353 Glen Cove Rd.
PO Box 803
Port Townsend, WA 98368
(360) 385-7258
www.nimbaanvils.com

Pieh Tool Company
661 E Howards Road, Suite J
Camp Verde, Arizona 86322
(888) 743-4866
www.piehtoolco.com

Timpercove Trading Co.
121 H St.
Petaluma, CA 94920

Coal

Bradford Coal Company, Inc.
PO Box 368
Bigler, PA 16825
(814) 342-0429
Contact: Terry Swanson
Blacksmith coal.

City Coal Yard
116 N. Depot St.
Brazil, IN 47834
(812) 448-0428
Excellent blacksmith coal. Pocahontas #3 washed from West Virginia.

Coalesce
PO Box 7701
South Lake Tahoe, CA 96158
(916) 544-0428
Source of excellent blacksmithing coal. High BTU value, low ash and sulphur. Cokes well.

Cumberland Elkhorn Coal and Coke
950 Swan St.
Louisville, KY 40204
(502) 589-5300
Source of excellent blacksmithing coal in bulk or by the bag. The coal is washed, is low in ash, and produces little clinker. They will ship by truck.

Haines Holt
Federalsburg Coal
323 Holt St.
Federalsburg, MD 21632
(410) 754-5244
Good-quality blacksmith coal from north-central Pennsylvania.

National Coal Company
273 SW Cutoff
Worcester, MA 01604
(508) 752-7468

Phoenix Coal Sales, Inc.
310 S. Scraper St.
PO Box 598
Vinita, OK 74301
(918) 256-7873
(918) 256-7874

Paul Pierce Autobody and Powderly Coal Company
1924 Jefferson Avenue SW
Birmingham, AL 35211
(205) 925-3454
(205) 925-6877
Source of blacksmith or metallurgical coal. High BTUs, low sulphur, and low ash. They will ship anywhere.

Reiss Coal Co.
111 W. Mason St.
Green Bay, WI 54303-1573
(920) 436-7600
Good blacksmithing coal, either by the bag or in bulk.

Skei Coal & Wood
639 Lincoln Way
PO Box 394
Ames, IA 50010
(515) 232-4474
Suppliers of blacksmith coal.

H.M. Stevens
10749 150th St.
Edmonton, AB
Canada
(403) 389-2137
Supplier of blacksmith coal.

Utah American Energy
375 S. Carbon Ave. Ste. 127
Price, UT 84501
(435) 613-0805
Very good source of coal, high heat, low sulphur. Very reasonable and priced in bulk. Excellent people to work with. I have purchased from these folks for the last seven years.

Power Hammers

Postville Power Hammer
Blanchardville, WI
(888) 535-6320
Contact: Bob Bergman

Centaur Forge
117 N. Spring St.
Burlington, WI 53105
(262) 763-9175
(800) 666-9175

Ironwood, LLC
10385 Long Rd.
Arlington, TN 38002
(901) 867-7300
www.powerhammers.com
Contact: Brian Russell

Jerry Kennedy
1010 Barley Ln.
Buckner, MO 64016
(816) 650-3922

Kayne & Son
100 Daniel Ridge Rd.
Candler, NC 28715-9434
(828) 667-8868

Little Giant
H. "Sid" Sudemeier
420 4th Corso
Nebraska City, NE 68410
(402) 873-6603
Great service for all Little Giant parts and hammers.

Pieh Tool Company
661 E Howards Road, Suite J
Camp Verde, Arizona 86322
(888) 743-4866
www.piehtoolco.com

Ozark School of Blacksmithing
H.C. 87, PO Box 5780
Potasi, MO 63664
(573) 438-4725

Steel

Architectural Iron Designs, Inc.
950 S. 2nd St.
Plainfield, NJ 07063
(800) 784-7444
www.archirondesign.com
Ornamental components.

Art & Metal Co.
243 Franklin St.
Hanson, MA 02341
(781) 294-4446

Atlas Metal Sales
1401 Umatilla St.
Denver, CO 80204
(800) 662-0143
www.atlasmetal.com
Supplier of silicon bronze.

Bay Shore Metals, Inc.
PO Box 882003
San Francisco, CA 94188-2003
Ornamental steel suppliers.

King Architectural Metals
9611 East R.L. Thornton Freeway (I-30)
Dallas, TX 75228
(800) 542-2379
www.kingmetals.com

Lawler Foundry Corp.
PO Box 320069
Birmingham, AL 35232-0069
(800) 624-9512
Ornamental components.

M.S.C. Industrial Supply
75 Maxess Rd.
Melville, NY 11747-9415
(800) 645-7270
www.mscdirect.com
Good supplier for high carbon and tool steels, grinders, you name it. Great prices and excellent service. I've dealt with them for more than twenty years. They have fifty-six locations throughout the United States, Puerto Rico, and Mexico.

Triple-S Steel
6000 Jensen Dr.
Houston, TX 77026
(713) 697-7105
Ornamental components and industrial suppliers.

TS Distributors
4404 Windfern Rd.
Houston, TX 77401
(800) 535-9842 or (800) 392-3655
www.tsdistributors.com

Organizations

Schools

Appalachian Center for Crafts
1560 Craft Center Dr.
Smithville, TN 37166
(615) 597-6801

***DeLaRonde Forge**
PO Box 1190
Mancos. CO 81328
(970) 533-7093
www.delarondeforge.com

The Forgery School of Blacksmithing
13 Imnaha Rd.
Tijeras, NM 87509
(505) 281-8080

Haystack Mt. School of Crafts
PO Box 518AR
Deer Isle, ME 04627
(207) 348-2306
www.haystack-mtn.org

John C. Campbell Folk School
One Folk School Road
Brasstown, NC 28902
(800) FOLK-SCH
www.folkschool.com

New England School of Metalwork
7 Albiston Way
Auburn, ME 04210
(888) 753-7502
www.newenglandschoolofmetalwork.com

***North House School**
PO Box 759
Grand Marais, MN 55604
(888) 387-9762
(218) 387-9762
www.northhouse.org

Ozark School of Blacksmithing
H.C. 87- PO Box 5780
Potasi, MO 63664
(573) 438-4725

Penland School of Crafts
PO Box 37
Penland, NC 28765
(828) 765-2359
www.penland.org

Peters Valley Craft Education Center
19 Kuhn Rd.
Layton, NJ
(973) 948-5200
www.pvcrafts.org

Pieh Tool Company
661 E Howards Road, Suite J
Camp Verde, Arizona 86322
(888) 743-4866
www.piehtoolco.com

Touchstone Center for Crafts
1049 Wharton Furnace Rd.
Farmington, PA 15437
(800) 721-0177
www.touchstonecrafts.com

***Turley Forge Blacksmithing School**
919-A Chicoma Vista
Santa Fe, NM 87507
(505) 471-8608
www.turleyforge.com

**These are schools I know of personally.*

Magazines/Periodicals

Anvil Magazine
www.anvilmag.com

Anvil's Ring
www.abana.org
Official quarterly publication of Artist-Blacksmith's Association of North America (ABANA).

Blacksmith's Gazette
950 S. Falcon Rd.
Camano Island, WA 98292
www.fholder.com/Blacksmithing/default.htm
Monthly.

Blacksmith's Journal
(800) 944-6134
www.blacksmithsjournal.com
Loads of information!

Metal Working
Lindsay Publications, Inc.
PO Box 538
Bradley, IL 60915-0538
(815) 935-5353
www.lindsaybks.com

Miscellaneous

Hartford Insurance Corporation
Industrial Coverage Corporation
3237 Rt. 112 Bldg. 6
Medford, NY 11763
(800) 242-9872
www.industrialcoverage.com/company.cfm
Insurance for self-employed blacksmiths.

The Broom Shop
PO Box 1182
Grants Pass, OR 97528
(541) 474-3575
www.broomshop.com
Custom brooms for fireplace sets.

Plug Depot
Casey Warner
(585) 637-9036
Ear protection supplier.

Suggested Reading

Andrews, Jack. *Edge of the Anvil.* Drexel Hill, PA: Skipjack Press, 1994.
Good basic book, but it offers a complicated, mystical approach to tempering.

Back to Basics. Pleasantville, NY: Reader's Digest, 1981.
A great book covering many facets of self-sufficiency, including blacksmithing.

Book of Buckskinning, Vol. IV. Texarkana, TX: Scurlock Publishing Inc., 1987.

Drew, James M. *Blacksmithing.* St. Paul, MN: Webb Book Publishing Co., 1935.
One of the old standards, though it is short on illustrations.

———. *Farm Blacksmithing.* Guilford, CT: Lyons Press, 2000. Originally published by Webb Press, 1901.
A good book for the homesteader, but again lacking sufficient illustrations.

Grant, Madison. *The Knife in Homespun America.* York, PA: Maple, 1984.
One of the best idea books for primitive knives available.

Kauffman, Henry J. *American Axes.* Brattleboro, VT: Stephen Green, 1871.
Very good idea book showing many styles of axes.

McNerney, Katheryn. *Antique Tools: Our American Heritage.* Paducah, KY: Collector Books, 1979.
A good reference and idea book for hand tools.

McRaven, Charles. *Country Blacksmithing.* New York, NY: Harper & Row, 1981.
Another good basic book, though weak in the area of tempering. Has some good ideas and projects for the homesteader.

Meek, James B. *Art of Engraving.* Montezuma, IA: F. Brownell & Son, 1973.
An excellent basic introduction to engraving.

Meilach, Dona. *Decorative Ironwork.* New York, NY: Crown Publishing, Inc., 1977.
A fantastic idea book of contemporary smiths and earlier works.

Minnis, Gordon. *American Primitive Knives, 1770-1870.* Bloomfield, Ontario, Canada: Museum Restoration Service, 1983.
Excellent idea book for primitive knives.

Peterson, Harold L. *American Indian Tomahawks.* Heye Foundation, New York, NY: Museum of American Indian, 1965.
Many tomahawks; well illustrated.

Richardson, M.T. *Practical Blacksmithing.* New York, NY: Weathervane Books, 1978.
Originally published in four volumes, 1889–1891. This is a great book for the traditionalist.

Russell, Carl P. *Firearms, Traps and Tools of the Mountain Men.* Albuquerque, NM: U. of New Mexico Press, 1967.
Excellent reference book.

Schwarzkopf, E. *Plain and Ornamental Forging.* New York, NY: John Wiley & Sons, 1930.
Excellent old-time book written by a master of the craft.

Simmons, Marc and Frank Turley. *Southwestern Colonial Ironwork.* Santa Fe, NM: Museum of New Mexico Press, 1980.
A great idea book of traditional Southwest Spanish ironwork, including tools, utensils, ornamental, horseshoeing, and history.

Tucker, Ted. *Practical Projects for the Blacksmith.* Lompoc, CA: Larson Publishing Co., 1980.
A lot of good basic projects. Weak on tempering.

Watson, Aldren A. *The Village Blacksmith.* New York, NY: Thomas Y. Crowell Co., 1968.
Some good general and historical information. Good illustrations.

Weygers, Alexander. *The Making of Tools.* Van Nostrand Reinhold Co., 1974.
One of the best books out there.

———. *The Modern Blacksmith.* Van Nostrand Reinhold Co., 1974.
One of the best books out there bar none. The basics are covered very well. Easy to follow, easy to understand.

———. *The Recycling, Use, & Repair of Tools.* Van Nostrand Reinhold Co., 1978.
Though nontraditional techniques are used, such as vise grips, wrenches, and torches, this book is loaded with good information.

Glossary of Terms

anneal: To soften by cooling slowly.

anvil: A heavy, shaped piece of steel for shaping hot iron with hammers.

anydrus borax: Substance used for a flux in forge welding.

butt weld: Welding two pieces together by forge welding the two ends together. Usually used for larger diameter shafts.

carbon steel (high-carbon steel): Steel with enough carbon to be hardened and tempered for use in tools.

clinkers: Impurities and residue left over after the coal and coke have been consumed. They will have the appearance of hard, glass-like clumps when the fire cools.

coke: Partially burned coal that creates the high heat in the center of the fire.

cold chisel: A chisel used to cut cold steel.

cross-peen hammer: A hammer whose blunt chisel face is perpendicular to the handle.

curling: A decorative touch normally done after the tip has been flattened or drawn to a point.

Damascus steel (pattern welded): Mild steel and high-carbon steel forge welded together. True damascus was done during the smelting process and was called Wootz Steel.

draw out: To stretch a piece of steel by heating and hammering.

drift out: To enlarge or shape a hole.

drift pin: Specially shaped tool for forming tool eyes such as hammers, axes, picks, and the like.

flaring: A method usually used in reference to tubing or pipe where the end is heated and pushed out to be wider than the body (flared).

flux: Substance, usually Borax in combination with other materials, such as sand and metal filings.

forge weld: Joining of two pieces of steel by fluxing, heating to a white semi-molten state, and hammering together.

fuller/fullering: A tool with a narrow rounded face used to thin, narrow, spread, or draw out hot iron. It may be used singularly or in pairs; that is a top and bottom fuller.

hardy: Vertical chisel set into the hardy hole in the anvil.

hardy hole: Square hole in the face of the anvil used for holding tools.

heat colors: Progressive colors that are seen in the metal as it is heated, from dull red to a white heat.

heat treating: The sequential steps of heating, hardening, and tempering steels for use in tools.

hot cutter: Chisel or mechanical device used to cut hot metal.

leg or stump vise: A vise with a long leg for added support and stability.

mild steel: Steel with very low carbon content that cannot be tempered.

normalize: Cooling metal in the air to its normal hardness.

oxidation scales: Thin flakes of oxidized metal caused by red-hot to hotter metal being exposed to air. In some cases with large pieces, this may cause you to lose 1/8 inch of stock.

poll: The back side of an axe used as a hammer.

pritchel hole: The round hole in the face of the anvil near the head.

punch (punching): A tool and method for making holes of various sizes and shapes in hot metal.

quench: The process of hardening by immersing hot metal in cold liquid.

rivet: A method of joining two pieces of metal by use of pins hammered flat on each end.

scarf (scarfing): Drawing two ends down prior to welding them together.

straight-peen hammer: A hammer whose blunt chisel face is parallel to the handle.

temper: Drawing the brittleness or hardness out of steel by controlling the heat after the piece is hardened.

tempering medium: The various liquids used for heat treating: water, salt water, or oil.

trip-hammer (power hammer): A mechanical hammer powered by water, compressed air, or electricity.

upset: Thickening of the stock by heating and hammering; the opposite of drawing out.

wrought or black iron: An old, fibrous, virtually carbon-free steel. It is hard to find anymore.

JOE DELARONDE began his blacksmithing career 35 years ago with an apprenticeship under a master German blacksmith. Following his three-year apprenticeship, Joe performed general blacksmithing, including wheelwright and plow lay work. He found his niche re-creating the tools and weapons of the early frontier. His works are in use around the globe by military personnel and living history enthusiasts as well as in private collections and museums in the United States, Canada, Mexico, and Europe. He continues to work at his shop in Mancos, Colorado, producing some of the finest tomahawks, axes, and knives available on the market today.

Made in the USA
Monee, IL
23 March 2021

63001359R00077